Collaborative Process Automation Systems

Collaborative Process Automation Systems

Martin Hollender

Collaborative Process Automation Systems

Copyright © 2010 by ISA—International Society of Automation

67 Alexander Drive
P.O. Box 12277
Research Triangle Park, NC 27709

All Rights Reserved.
Printed in the United States of America.
10 9 8 7

ISBN: 978-1-936007-10-3

No part of this work may be reproduced, stored in a retrieval system, or transmitted in any form or by any means, electronic, mechanical, photocopying, recording or otherwise, without the prior written permission of the publisher.

Notice
The information presented in this publication is for the general education of the reader. Because neither the author nor the publisher have any control over the use of the information by the reader, both the author and the publisher disclaim any and all liability of any kind arising out of such use. The reader is expected to exercise sound professional judgment in using any of the information presented in a particular application. Additionally, neither the author nor the publisher have investigated or considered the affect of any patents on the ability of the reader to use any of the information in a particular application. The reader is responsible for reviewing any possible patents that may affect any particular use of the information presented.
Any references to commercial products in the work are cited as examples only. Neither the author nor the publisher endorses any referenced commercial product. Any trademarks or trade names referenced belong to the respective owner of the mark or name. Neither the author nor the publisher makes any representation regarding the availability of any referenced commercial product at any time. The manufacturer's instructions on use of any commercial product must be followed at all times, even if in conflict with the information in this publication.

Library of Congress Cataloging-in-Publication Data

Hollender, Martin.
  Collaborative process automation systems / Martin Hollender.
     p. cm.
  Includes index.
  ISBN 978-1-936007-10-3 (pbk.)
 1. Process control. 2. Automatic control. 3. Logic design. 4. Control theory. I. Title.
  TS156.8.H655 2010
  670.42'7—dc22
                          2009036018

To Sigrid and Anton

Table of Contents

Preface   ix
Acknowledgments   xi

Chapter 1

1.1   Introduction 2
1.2   Industries and Utilities 9
1.3   The History of Process Automation Systems 18
1.4   The ARC Advisory Group's Collaborative Process
      Automation System (CPAS) Vision 25

Chapter 2

2.1   CPAS System Architecture 42
2.2   Common Object Model 56
2.3   Industry Standards and Open Interfaces 67
2.4   OPC Unified Architecture 86
2.5   Dependable Automation 100
2.6   Scalability and Versatility 112
2.7   IT Security for Automation Systems 118
2.8   Safety—One of the Key Practices in Control Engineering ... 133
2.9   Traceability 147

Chapter 3

3.1 Engineering 156

3.2 Control Logic Programming 183

3.3 CPAS Functional Units 191

Chapter 4

4.1 Alarms and Events 210

4.2 Common Process Control Application Structure 224

4.3 Remote Operation and Service 238

Chapter 5

5.1 Advanced Process Control 244

5.2 Loop Tuning 258

5.3 Loop Monitoring 269

5.4 Plant Asset Management 281

5.5 Information Management 294

5.6 Enterprise Connectivity 313

5.7 Planning and Scheduling 330

Chapter 6

6.1 Operator Effectiveness 352

6.2 Alarm Management 378

List of Acronyms   393

Index   399

Preface

There are many books about control systems on the market, but a gap in this area still remains. This book is intended to fill that gap.

Most books on control systems present control theory. They mainly dwell on the theory behind the analysis and synthesis of continuous closed control loops, but they may also present methodologies for the control of discrete event systems. Students, both undergraduate and graduate, have a variety of choices for learning how to design control systems.

Some books deal with control technology. They provide information about the inner structure and functioning of control processors and about the communication technologies used to combine controllers. Both for students and for practitioners, they describe how to develop communication stacks for fieldbuses and how to program controllers.

All these books have their merits, but they fail to convey an impression of what industrial control systems are really about. Beyond the basic control functions, modern process control systems present a wealth of further functions which allow the safe, secure, and efficient operation of an automated plant and which deliver significant benefits to the plant owner and operator. Furthermore, these functions and how they are integrated with each other and with the enterprise resource planning (ERP) system tend to become the differentiating factors between the process automation systems on the market. Therefore, automation system suppliers spend significant efforts on these topics, and their endeavors deserve attention. For engineering and computer science students, modern process control systems offer a variety of interesting topics to delve into, and valuable career perspectives afterwards. Engineers, both the novice and the experienced, should familiarize themselves with these developments to be able to understand modern process control system functionality and to make the best use of it.

This is the aim of this book. It not only conveys the overall picture of modern process control systems, but provides a thorough understanding of the functions that turned the "classical" distributed control system (DCS) into the so-called "Collaborative Process Automation System (CPAS)." It does not propose to deal with all the details but to stimulate further reading, for which it provides appropriate hints. Thus, this book is unique in focus and structure. I wish for this book the numerous interested readers that it deserves.

—Prof. Dr.-Ing. Alexander Fay
Helmut-Schmidt-University, Hamburg, Germany

Acknowledgments

The contributors who have written or reviewed chapters for this book come from many different countries. I have always enjoyed this kind of international cooperation throughout my career. I'm very grateful to all colleagues who directly or indirectly contributed to this book and am especially honored that Dave Woll from ARC, one of the creators of the CPAS concept, has contributed a section defining the ARC CPAS vision.

The following persons were of invaluable assistance by reviewing one or more of the chapters: Margret Bauer, Carlos Bilich, Armin Boss, Rainer Drath, Krister Forsman, Christopher Ganz, Georg Gutermuth, Stefan Heiß, Alexander Horch, Gunnar Johannsen, Jim Kline, Michael Lundh, Dejan Milenovic, Per Erik Modén, Mikael Rudin, Wolfgang Schellhammer, Mario Schwartz, Rob Turner, and Kay Wilke. Many thanks also to my wife, Sigrid Hollender, who helped with language issues.

This book was made possible by the outstanding contributions of Susan Colwell (ISA) and Scott Bogue (Words$_2$Work). They are to be recognized for their exceptional efforts in making this book a high-quality piece of work.

Many thanks to the following authors, who contributed insightful and well-reasoned chapters in their areas of expertise:

Chapter	Author
The ARC CPAS Vision	Dave Woll
OPC UA	Wolfgang Mahnke
Dependability	Hubert Kirrmann
Security	Martin Naedele/Markus Brändle
Safety	Zaijun Hu
Engineering	Georg Gutermuth
Control Logic Programming	Georg Gutermuth
CPAS Functional Units	Christopher Ganz

Common Process Control Application Structure	David Huffman
Advanced Process Control	Alf Isaksson
Loop Tuning	Alf Isaksson
Loop Monitoring	Alexander Horch
Plant Asset Management	Alexander Horch/Margret Bauer
Enterprise Connectivity	Margret Bauer/Sascha Stoeter
Planning and Scheduling	Iiro Harjunkoski
Operator Effectiveness	Tony Atkinson/Martin Hollender

CHAPTER 1

1.1 Introduction

Martin Hollender

The idea for this book came into being during a discussion with Martin Naedele, because we both believed there is a need for a book that makes the CPAS concepts concrete using an existing system and its features. The knowledge necessary to automate a modern industrial plant is so broad that no single engineer can have it all. Such automation projects are usually done by a team of highly specialized engineers. Similarly, a book on CPAS can not be written by a single author. Many existing books on process automation are thick and heavy; for example, the "Handbuch für Prozessautomation" (Früh et al., 2008) includes more than eight hundred pages, and the "Instrument Engineers' Handbook" (Lipták, Ed., 1995) comes in three volumes totaling several thousand pages. Many existing books focus on control theory. Other books focus on hardware, sensors, and actuators. Very few books, however, span all the important parts of process automation necessary to reach world-class automation. Areas already covered in many other books, such as digital fieldbuses, control theory, or IEC 61131 programming, are covered in less depth in this book and pointers are given to the relevant literature.

The concept for this book is to put modern and important topics, which are often difficult to find in other literature, into a single handy book. Topics cover a well-established state of the art in industrial plants. There is one exception: The current state of the art for OPC are the "classic" standards DA, AE, and HDA. For the brand-new OPC UA standard only a few first products exist. But as I'm deeply convinced that OPC UA will play a central role in future CPAS systems, an extra chapter was added on OPC UA.

This book aims to be abstract, general, and relatively vendor neutral, but many of the examples are explained using ABB's System 800xA. As Dave Woll writes in his chapter, the term CPAS (Collaborative Process Automation Systems) as defined by ARC (Automation Research Corporation) does not describe a particular commercially available system. ARC's CPAS definition is an excellent guiding framework covering the essential aspects of modern process automation.

1.1 Introduction

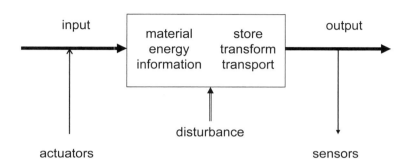

Figure 1–1 Technical Process

Overview

The automation of industrial processes supports the storage, transportation, and transformation of raw materials into useful products (see Figure 1–1). Processes can be classified into batch and continuous processes. Batch processes are characterized by the production of a given amount of product in batches (e.g., pills, cakes, or beer). Continuous processes transform raw materials into finished or semi-finished products, usually in a steady-state process. A typical example is refining, where crude oil is refined into gasoline in a continuous process that often runs for years without any interruption. Examples of the process industries include food and beverages, chemicals, pharmaceuticals, petroleum, metals, pulp and paper, cement, and (although the "product" is not physical), power generation. Sensors (e.g., flowmeters or thermometers) are used to get information about the process, and actuators (e.g., valves, pumps, or heat exchangers) can be used to influence the process.

Related subject areas like the automation of discrete (piece) manufacturing, buildings, airplanes, and trains have much in common with process automation but are outside the scope of this book.

Process automation is often drawn in automation pyramids with several automation layers (see Figure 1–2). Such pyramids correlate with the management hierarchy of a company and show how information is more and more condensed. On the lowest level the system deals with physical values like pressure, flow and temperature whereas higher levels deal with concepts like product quality, production schedule, and profitability. An example are thousands of measurements along a paper reel condensed into a single quality parameter determining the price of the paper reel. The real-time requirements are strictest on the lowest level, where the automation system needs to guarantee response times in the range of seconds and sometimes milliseconds, and relax toward the higher levels, where the time horizon may extend to weeks. The crucial fact is that several different layers with different requirements exist and need to interact with each other.

The ISA-95 series of standards (ISA, 2007) contain the most recognized definition of the different automation layers (Figure 1–3) and are discussed in detail in Section 5.6. In many application areas the control algorithms are mature, and relatively simple PID (proportional-integral-derivative) controllers are sufficient for most of the tasks. While control at the levels 2,1,0 is quite well mastered, the integration of the higher levels now plays a more and more important role in the optimization of production.

The biggest change in future plant performance improvement will come from the empowerment of the operator (Woll et al., 2002). Production will undergo fundamental organizational changes because operators are becoming knowledge workers empowered with information. This proliferation of information is allowing organizational structures to flatten, shifting the authority and responsibility associated with the distribution of information down to lower levels. A higher degree of coordination at lower levels is therefore required.

Another important factor in plant performance improvement is providing a unified platform for plant and maintenance management that also automates the transactions between the automation system and computerized maintenance management systems (CMMS).

As the capabilities of traditional distributed control systems (DCSs) are no longer sufficient to cover all relevant areas of process automation, a new generation of systems was created. The concept of Collaborative Process Automation Systems (CPAS) was developed by Automation Research Corporation (ARC) (www.arcweb.com), a consulting company based in Boston, Massachusetts. A key aspect of a CPAS is the ability to present information in context to the right people at the right time from any point within the system, to include a single, unified environment for the presentation of information to the operator. In addition, a key strength of a CPAS is the ability to extend its reach beyond

1.1 Introduction

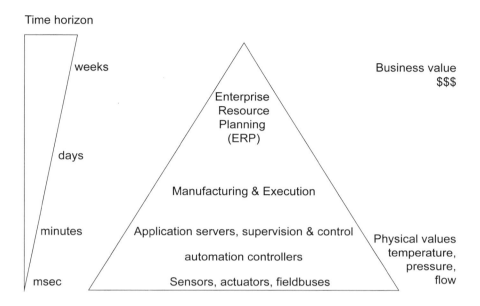

Figure 1–2   Automation Pyramid

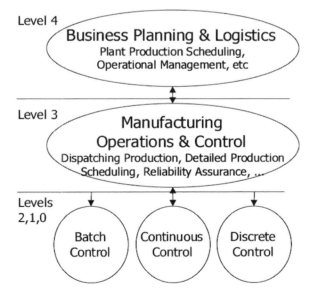

Figure 1–3   Levels of an ISA-95 Functional Hierarchy Model

the traditional capabilities of the DCS to include functions such as production management, safety and production-critical control, advanced control, information management, smart instrumentation, smart drives and motor control centers, asset management, and documentation management capabilities.[1]

Safety and control are linked and embedded within the same architecture, providing a common high integrity system environment for production control, safety supervision, and production monitoring. This architecture provides the option of combining control and safety functions within the same controller or keeping control and safety functions separate within the same system. With safety and process applications executing within the same system environment, and even within the same controller, a CPAS can offer safe, instant interaction between applications. Key features of a CPAS include:

- Extensible
- Common Data Model
- Adaptable through Configuration
- Single Version of the Truth
- Interoperable
- Standards Adoption
- Multi-supplier Support
- Security and Reliability
- Actionable Context Support
- Knowledge Workplace Support

Multiple software providers can manage parts of the system's functionality, but no single software vendor is capable of managing the entire scope of a CPAS. Therefore, application architecture design is crucial for successful projects.

The most important trends that promote change in automation development (Jämsä-Jounela, 2007) include globalization, networked economy, and sustainable development.

---

1. Smart instruments include a microprocessor and have much more functionality than just measuring a value. They can include functionality for device management, calibration and diagnosis.

Structure of the Book

Chapter 1 starts with a general introduction to process automation, represented industries, and historical developments, allowing readers new to the field to build up some context. Next the CPAS concept is defined and explained in detail.

Chapter 2 describes the infrastructure required to build up a CPAS. The chapter discusses the specific requirements that drive the architecture of a CPAS. For example, IT (Information Technology) security was hardly an issue 20 years ago, but in today's networked economy with its high danger of cyber-terrorism, IT security has become a top priority that needs to be built into a CPAS from the ground up. It also explains why standards are essential for CPAS and introduces the most important standards. It is surprising that many engineers have never heard about important base standards such as XML or OPC. The quality of many automation solutions depends on a thorough understanding of these base standards and the standards that build on them. This book cannot cover each standard in depth, but it tries to help the reader understand their importance and to stimulate further reading.

Chapter 3 explains the engineering of a CPAS. As the cost for hardware has decreased and system complexity has increased, the total cost of a solution is now largely determined by software engineering efficiency. The reuse of best practice solutions in libraries plays a key role.

Chapter 4 covers topics on the lower levels of the automation pyramid. It starts with the description of an Alarms & Events subsystem. In the chapter on the Common Process Control Application Structure, Dave Huffman explains why many principles developed to control batch processes also have benefits for other processes. The chapter concludes with a discussion of control system remote operation and service.

Chapter 5 covers higher level topics (from an automation pyramid point of view) such as information management, enterprise connectivity, and advanced process control.

Chapter 6 discusses operator effectiveness and the role of human operators in highly automated systems. Knowledge about human-machine systems helps engineers to design better human system interfaces (HSI), which results in better overall system performance. The book concludes with a chapter on alarm management.

References

Früh, K. F., Maier, U., and Schaudel, D. *Handbuch der Prozeßautomatisierung*. München: Oldenbourg Industrieverlag, 2008.

International Society of Automation (ISA). *ANSI/ISA-95.00.01-2000, Enterprise-Control System Integration — Part 1–5*, 2007.

Jämsä-Jounela, S. L. "Future Trends in Process Automation." *Annual Reviews* in *Control,* No. 31 (2007) pp 211–220.

Lipták, B. G. (Ed.) *Instrument Engineers' Handbook Vol. I–III*. London, New York et al.: CRC Press, 1995.

Woll, D., Caro, D., and Hill, D. *Collaborative Process Automation Systems of the Future*. Boston: Automation Research Corporation; arcweb.com, 2002.

1.2 Industries and Utilities

Martin Hollender

Introduction

The focus of this book is the automation of industrial processes. Many of these processes transform raw materials (like wood, iron ore, or crude oil) into refined products (like paper, steel, or gasoline). Usually this is done by continuous production on a large scale and these industries are sometimes called heavy industries. Often streams of materials are flowing 24 hours per day. Another type of process is a batch process where ingredients are usually processed in vessels and then moved to the next production phase like in the pharmaceutical or the food and beverages industry. Crude oil production and mining are examples of non-transformative transport processes, which also are controlled by a CPAS.

Automation of industrial processes usually requires lots of continuous signals and closed-loop control. The process models are often complex, based, for example, on non-linear differential equations. In many cases sufficiently precise process models cannot be obtained with reasonable effort, and therefore closed-loop control needs to be applied.

Manufacturing processes like the production of automobiles or electronic devices are concerned with discrete items and can normally be precisely modeled, for example with Petri nets. Automation of discrete manufacturing processes is usually done with binary signals and open loop control.

Although the automation of both areas has many common requirements, there are important differences, justifying a focus on only one of the areas. The remainder of this book will focus on the automation of continuous industrial processes, with attention being paid to batch (discontinuous) processes where commonality exists.

Figure 1–4 shows how the worldwide process automation market is divided into different industries. There is a total $32.8 billion market (2007) for automation systems for continuous and batch process industries worldwide. The following section gives a short overview of the specifics in several key industries.

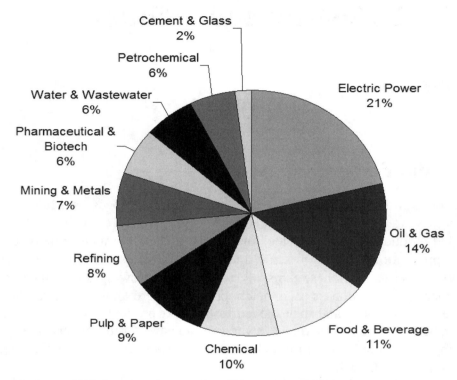

Figure 1–4   ARC Process Automation Shipments to Different Industries (Shah, 2007)

Electric Power

Power plants transform fossil fuels, nuclear fuels, or renewable energy like wind or water into electrical energy. Modern, low-emission coal-fired power stations will form the backbone of the world's electricity supply in the foreseeable future. A typical 770 megawatt (MW) plant burns 120 lb (55 kg) of coal per second at a temperature of 2460°F (1350°C). The most efficient stations can achieve an efficiency of up to 50 percent. Such plants fulfill various expectations; on the one hand, the security of supply (i.e., power is available at all times). This allows the performance characteristics of intermittent, renewable energy sources to be compensated for. On the other hand, expectations regarding energy efficiency are fulfilled. The CO_2 emissions of these plants are almost 30 percent lower than those of older plants. Such efficient plants, however, place higher demands on automation and optimization technology. This results from the fact that their design and operation push modern materials almost to their limits. At the same time, the demands on transient behavior (changes in power output) with respect to adaptability and control capabilities in a changed power distribution network have increased. This network is characterized by power trading and by greater dynamics due to the increasing share of renewable energy.

More information on electric power generation may be found at:

- www.epri.com—The Electric Power Research Institute
- www.vgb.org—VGB Power Tech e.V.

Oil and Gas

The American Petroleum Institute divides the petroleum industry into the following sectors:

- Upstream (exploration, development, and production of crude oil or natural gas). Typical characteristics include adverse environments, extreme climates and limited space. More and more offshore operations are supervised remotely from onshore operation centers. Important requirements are high safety, maximum availability, extreme ruggedness and compact, modular designs. Rising oil prices make new methods and technologies interesting, like the processing of oil sands.

- Downstream (oil tankers, refiners, retailers, and consumers). Refineries convert crude oil into components like gasoline, diesel oil, kerosene, heating fuel oils and asphalt. There is a worldwide shortage of refinery capacity. Many refineries produce at maximum capacity without any interruption for long periods of time. A larger refinery can produce 80 gallons (300 liters) of gasoline per second. CPAS help to boost operating efficiency and product yields and to minimize emissions and waste.

- Pipelines move crude oil from wells on land and platforms in the oceans to refineries, and then to terminals where fuels are released to retail outlets.

- Marine. The Marine Segment involves all aspects of transporting petroleum and petroleum products by water, including port operations, maritime firefighting and oil spill response. Oil tankers make up a major portion of this segment.

Petroleum is also the raw material for many chemical products, including pharmaceuticals, solvents, fertilizers, pesticides, and plastics. Liquefied natural gas (LNG) is natural gas that has been converted to liquid form for ease of storage or transport. In these industries, CPAS helps to optimize operations, eliminate losses, improve the quality and flow of information, and meet safety and environmental regulations.

More information on the petroleum industry may be found at:

- www.api.org—American Petroleum Institute

Food and Beverage

Food and beverage processing includes techniques like fermentation (e.g., in beer breweries, peeling, spray drying, pasteurization, boiling, and frying). Food and beverages are mostly produced in batches. Processed foods are better suited for long distance transportation from the source to the consumer.

Food safety is the top issue. The three major regulatory areas most affecting the food and beverage industry are food safety tracking and tracing requirements, truth in labeling requirements, and environmental air-water-waste requirements. The food and beverage manufacturing industry is one of the largest consumers of energy and water and one of the largest producers of water and land environmental waste products (Clouther and Blanchard, 2007). Today, the industry is focusing on global super-branding, expansion of distribution channels, and additional process automation and integration of manufacturing, warehouse, logistics, and business systems.

For more information, see:

- foodindustrycenter.umn.edu—The Food Industry Center at the University of Minnesota
- www.ift.org—Institute of Food Technologists

Pharmaceuticals

The worldwide market 2007 for pharmaceuticals was $712 billion (VFA, 2007). Medical drugs are usually produced in batch processes, often with biotechnology. The regulatory constraints for drug production are massive to ensure patient safety and have a huge impact on CPAS (see Section 2.9).

For more information, see:

- www.ifpma.org—International Federation of Pharmaceutical Manufacturers and Associations (IFPMA)
- www.vfa.de—Association of Research-Based Pharmaceutical Companies
- www.fda.gov—US Food and Drug Administration

Chemical

In 2007, turnover for the global chemical industry was estimated to reach $3180 billion (ICCA, 2008). Petrochemicals initiated the plastics and materials revolution; fine and specialty chemicals offer a multitude of products, both for consumer items and industrial applications or processes, including active ingredients for crop protection and intermediates for pharmaceuticals. Synthetic dyes are central to the production of textiles. Industrialized countries account for a major part of world production, but the main growth centers of chemical sales and production are in emerging Asia. Figure 1–5 shows the distribution of different chemical products per the example of the German Chemical Industry.

Leading chemical manufacturers are coming up with effective strategies in today's highly competitive global marketplace. They include:

- Offering new and improved products
- Offering quality products
- Offering customized products
- Improving customer service
- Establishing large integrated manufacturing sites

In addition to the above strategies, chemical manufacturers emphasize their supply chain and business processes. These companies need to improve their manufacturing processes, which include better engineering, improved

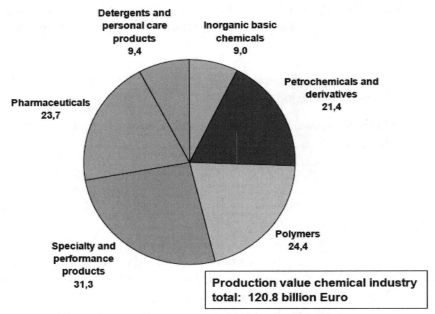

Figure 1–5   Chemical Products Per the Example of the German Chemical Industry

automation, closer integration of manufacturing and business systems, and better utilization of manufacturing assets (Ghosh, 2007).

More information may be found at:

- www.vci.de—Verband der Chemischen Industrie
- www.namur.de—International User Association Of Automation Technology In Process Industries
- www.icca-chem.org—The International Council of Chemical Associations

Pulp and Paper

Paper mills transform wood and recycled paper into large reels of paper. As producing paper is energy intensive (the production of one ton of pulp requires approximately 700 kWh), energy efficiency is a major issue for paper mills. The profitability of individual pulp and paper companies depends largely on production efficiency. Paper production integrates continuous and batch production (paper reels). Many different qualities can be produced on one single machine. Losses because of grade changes need to be minimized.

Technology trends in the pulp and paper industry include:

- Embedded control of machinery
- Advanced sensors
    - data analysis
    - fault diagnosis
- Advanced control systems
    - multi-variable controls
    - predictive control
    - predictive optimization
- Decision support tools
    - optimization of process, costs, quality, runnability, etc.
    - multi-criteria optimization

More information is available at:

- www.tappi.org—Technical Association of the Pulp and Paper Industry
- www.cepi.org—Confederation of European Paper Industries
- www.afandpa.org—American Forest & Paper Association

Mining and Metals

Materials recovered by mining include base metals, precious metals, iron, uranium, coal, diamonds, limestone, oil shale, rock salt, and potash. Surface (open pit) mining is the predominant exploitation procedure worldwide. In open pit mining, a mechanical extraction method, a thick deposit is generally mined in benches or steps.

CPAS provides central monitoring, operation, and control of the systems components and complex operational procedures of a mine. Process data, radio signals, video images, and satellite messages that are often gathered miles away are recorded and processed for this purpose. Automation-supported equipment includes excavators, spreaders, belt-conveyor systems, belt wagons, stackers-reclaimers, crushers and shovel excavators. Mineral processing includes the pelletizing and direct reduction of iron ore and the smelting and refining of non-ferrous metals. CPAS can optimize production through better average grind size, lower specific energy cost, increased reagent efficiency, and improved recovery. Important plant operational information such as metallurgical accounting and mass balancing are important for efficient energy management and optimization.

For more information, see:

- www.tms.org—The Minerals, Metals & Materials Society

Water and Wastewater

The water industry provides drinking water and wastewater services. Stringent regulations drive investments in CPAS solutions in the water and wastewater sector. One of the key challenges that CPAS manufacturers are facing is the need to provide systems that seamlessly integrate with existing plant infrastructure. These manufacturers will need to provide automation and control systems that are compatible with the existing plant set-up and are easy to integrate. End users are conservative in their outlook on the question of revamping existing systems in order to incorporate the latest automation and control solutions. Integration issues and the high costs involved in adopting new automation and control solutions are the key factors responsible for end-user reluctance. This has either prohibited or delayed the implementation of new automation and control solutions.

Additional information on this industry may be found at:

- www.watersupply.com

Cement

Cement manufacturing focuses on producing consistent, high-quality cement, meeting environmental responsibilities, improving process control, and

1.2 Industries and Utilities 17

managing information effectively to reduce capital employed. The use of CPAS minimizes material and energy use in complex processes. Each year, environmental regulations get tougher, and the more stringent the regulations become, the more money is spent on waste processing and disposal. As emissions from cement production decrease to all-time lows, cement kilns are recognized as excellent incinerators of waste materials; however, product and operating standards must remain high. Proactive, flexible, and adaptive systems are key.

For more information, see:

- www.cement.org—PCA Portland Cement Association

References

Clouther, S. and Blanchard, J. *Food & Beverage Industry Trends, Challenges and Opportunities*, ARC Strategies, www.ARCweb.com, 2007.

Ghosh, A. *Chemical Industry Trends, Challenges and Opportunities*, ARC Strategies, www.ARCweb.com, 2007.

International Council of Chemical Associations (ICCA) *REVIEW 2007–08* available at http://bit.ly/FGUNU (last retrieved 2 May 2009).

Schah, H. *Automation Systems for Process Industries Worldwide Outlook*, www.ARCweb.com, 2007.

1.3 The History of Process Automation Systems

Martin Hollender

Introduction

Historically, the production of goods like paper or iron was done manually by craftsmen and involved little or no automation. Later on, wind or water energy was translated into rotational energy, which was the first phase of automation. Process equipment like valves had to be operated manually and instruments such as thermometers had to be directly read by the operators.

The invention of the steam engine in the eighteenth century was an important basis for industrialization. Early plants often included sophisticated mechanical control schemes. Later on, pneumatic and electronic regulating controllers were introduced. The Jacquard Loom was invented around 1800 (see Figure 1–6). It used punch cards to control the design of the produced textiles. The link to modern computer-controlled batch recipes is clearly visible. After the invention of the computer, computer-based control systems were introduced in the 1950s.

Computerized automation controllers were developed from three different branches. In the automobile/discrete manufacturing industry, the programmable logic controller (PLC) was introduced, whereas in the petrochemical/process industry the term distributed control system (DCS) was established. Both had much in common but as the requirements for the industries are different, the solutions historically were different. For geographically distributed processes like pipelines, electrical power, and gas or water networks the automation was called supervisory control and data acquisition (SCADA). Today, as computer and network technology have rapidly matured, the borderlines between the areas is getting blurred. Each area has its specific requirements, but all share a huge set of common requirements that can be supported by the same technology.

1.3   The History of Process Automation Systems   19

Figure 1–6   Jacquard Loom *(Photo by Jeremy Keith)*

Programmable Logic Controller (PLC)

In 1968, General Motors requested an electronic replacement for electromechanical relay logic circuits, which were then in common use in automation. In response to this request, the company Modicon (Modular Digital Controller) developed the first PLC.

One of the languages used to program a PLC is called *ladder logic*, which directly tries to emulate the old relay-based ladder logic structures.

A modular PLC (Figure 1–7) combines CPU modules with a flexible number of analog or digital input/output (I/O) modules via a backplane bus. A PLC is designed for multiple input and output arrangements, extended temperature ranges, immunity to electrical noise, and resistance to vibration (Figure 1–8). Programs to control machine operation are typically stored in battery-backed or non-volatile memory.

At the core of a PLC are one or several central processor units (CPUs). The PLC gets the process measurements via digital or analog input cards. The results of the control logic are then written to digital or analog output cards. An alternative is to integrate digital fieldbuses. Depending on the size of the process, more or fewer I/O cards are required. A PLC does not include a keyboard,

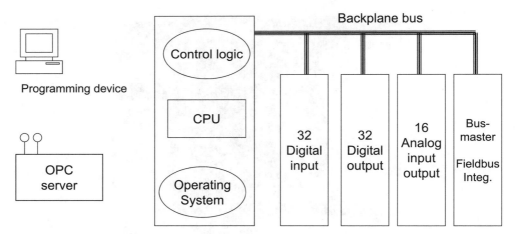

Figure 1–7 Modular PLC

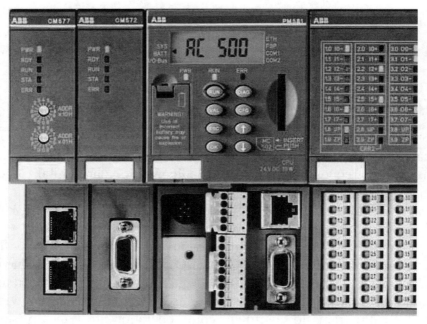

Figure 1–8 Modern PLC

hard disk drive, or mouse because it is designed to run under harsh conditions. The control logic software is usually developed on a separate programming device, in most cases a laptop. This software is then uploaded to the PLC.

Each cycle of a PLC program (Figure 1–9) starts with reading the input values from the input cards into memory. In the following main step, the con-

1.3 The History of Process Automation Systems 21

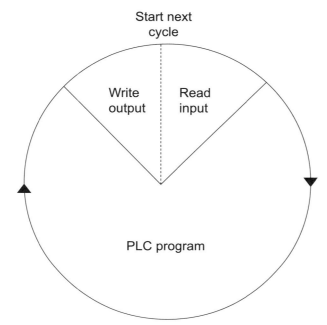

Figure 1–9   Cyclical PLC Execution

trol logic program is executed. Finally, the values of the output variables are written to the output cards. Required cycle times differ between industries, but one second is a typical cycle time. If more program steps need to execute than fit into one cycle, additional or faster hardware modules need to be introduced.

The multi-vendor non-profit association PLCopen (http://www.plcopen.org) promotes the development of compatible software for PLCs. The main focus of PLCopen is the IEC 1131-3 Standard (see Sections 2.3 and 3.2).

Distributed Control Systems (DCSs)

The first industrial control computer system was that of the Texaco Company at its Port Arthur, Texas, refinery with an RW-300 of the Ramo-Wooldridge Company; closed-loop control was achieved on March 15, 1959 (Stout and Williams 1995). The first serious proposal for direct digital control of full-size chemical plants, i.e., without the use of an intermediate analog, electronic, or pneumatic control system, was made by ICI in the early 1960s. Since the mid-1970s, microprocessor-based distributed control systems (DCSs) have become more and more standard for industrial continuous process control.

The early DCSs were closed, monolithic systems from a single vendor; that is rarely the case today.

In addition to the automation controllers, a DCS includes application servers such as process historians and the workstations, networks, and other modules necessary to build a complete system (see Figure 1–10).

Rapid advances in hardware and software technology have blurred the difference between the systems described as DCS and PLC (see Figure 1–11). For example, IEC 1131-3 programming, which originates from the PLC side, is also now standard for a DCS.

Supervisory Control and Data Acquisition (SCADA)

Supervisory control and data acquisition systems are used for process control in real time for geographically distributed processes like pipelines, electrical power distribution, and water or gas networks. SCADA is also used in power plants, as well as in oil and gas refining, telecommunications, transportation, and water and waste control.

Historically, SCADA systems collected data from remote terminal units (RTUs), often over very low-bandwidth connections like telephone lines, and conveyed simple instructions from a master terminal unit (MTU) to the RTUs. As reliable and fast communication infrastructure, for example, via radio or satellite gets more and more commonplace, geographical distance requires less and less a special architecture. In the future the differentiation between PLC, DCS, and SCADA will make less and less sense. As we have seen, each of them has its historical roots in different domains but the general requirements that must be satisfied for each get more and more similar.

Collaborative Process Automation Systems (CPAS)

Today, plants ask for systems offering more than just distributed control. These systems must be able to integrate all kinds of relevant applications like Advanced Process Control or Information Management and to support a high level of collaboration. The CPAS concept is described in more detail in Section 1.4 "CPAS", by Dave Woll and in the remainder of this book.

1.3 The History of Process Automation Systems

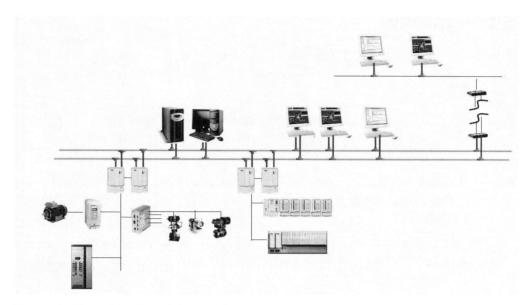

Figure 1–10   Distributed Control System

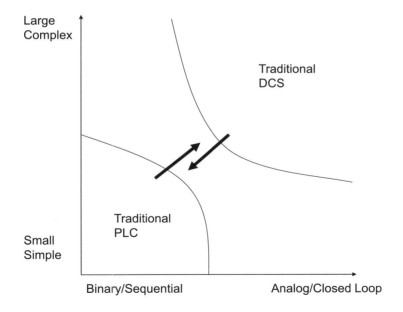

Figure 1–11   The Historically Separate Concepts of PLC and DCS Are Approaching Each Other

Bibliography

O'Brien, L. and Woll, D. "DCS Marks 30 Year Journey to Operational Excellence." *ARC Insights Report 2005-36MP*, www.arcweb.com, 2005.

Rarnebäck, C. "Process Automation Systems—History and Future." *IEEE Conference on Emerging Technologies and Factory Automation,* Lisbon, Portugal, 16–19 September 2003: 3–4.

Samad, T., McLaughlin, P., and Lu, J. "System Architecture for Process Automation: Review and Trends." *Journal of Process Control 17*, 2007: 191–201.

Stout, T. M. and Williams, T. J. "Pioneering Work in the Field of Computer Process Control." *IEEE Annals of the History of Computing,* Volume 17, Issue 1, 1995.

1.4 The ARC Advisory Group's Collaborative Process Automation System (CPAS) Vision

Dave Woll

Introduction

Collaborative Process Automation System (CPAS) is the ARC Advisory Group's vision of how process automation systems should evolve through the current decade. It does not describe a particular commercially available system. The technology used in CPAS is available and proven; it is not all available in any particular offering but is available across all systems.

Primarily, CPAS should be considered an environment that enables applications including Process Control, Advanced Process Control, and Operations Management complemented by human empowerment applications such as decision support and advanced analytics.

CPAS spans from sensors and actuators to ERP interfaces, and for the sake of clarity and minimized complication, it does not recognize software such as MES or HMI SCADA as subsystems but rather addresses these as classes of applications. Within the constraints of the IEC 61131-3 programming and configuration standard, it incorporates a single model with distributed processing. It is data driven, all digital and based on true international standards. CPAS is inherently robust, delivers high accuracy with low Total Cost of Ownership (TCO), and supports a high level of collaboration.

The Lineage of CPAS

CPAS is the product of several decades of technological process automation evolution (Figure 1–12), each with its specific characteristics and focus.

The Computer Integrated Manufacturing Decade

The CPAS Vision discussion begins with the introduction of "Computer Integrated Manufacturing." This was ahead of its time and was largely an academic exercise because computer and network technology was not in place to support it. It was focused on supervisory applications and its primary value was to validate what was possible and to set the stage for future developments.

The System (DCS) Centric Decade

This decade brought us the microprocessor and distributed process automation in the form of the first distributed control systems (DCSs). These marked the transition from analog to the precision of digital technology and utilized IEEE 802.4 token-passing communications because of its determinism. These systems were essentially proprietary in nature, with closed communications.

The Network Centric Open Systems Decade

The next decade began to address the proprietary nature of systems and brought us the "Network Centric Open System," with much of the associated technology developed by the U.S. Department of Defense (DoD). The DoD developed the first open standards-based operating system, Unix, and more important, a truly state-of-the-art standards-based networking technology, IEEE 802.3 Ethernet TCP-IP. Rather than utilize token passing, it utilizes collision algorithms. This networking technology enabled client-server computing, the Internet and ultimately CPAS with global data access. The majority of the process automation system suppliers were late in embracing Ethernet because they assumed it was not deterministic enough; however, it was later proven that this was not the case. This decade also saw the introduction of the first commercially available Process Data Historians and because most process control systems had not embraced Ethernet, most systems became historian-centric with respect to data access for supervisory applications. Ethernet TCP-IP has

1.4 The ARC Advisory Group's Collaborative Process Automation System (CPAS) Vision

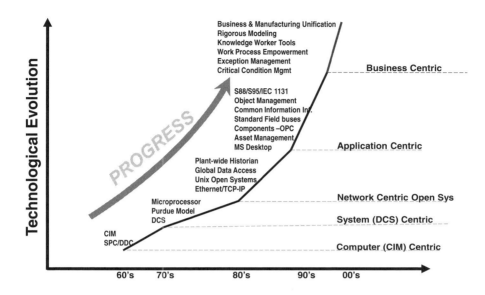

Figure 1–12 The Evolution of Automation Systems

become ubiquitous as the state-of-the-art networking infrastructure for process automation and enterprise systems.

The Application Centric Decade

A common foundation for networking made it possible to incorporate process control and information into the same application environment and thereby made the next decade, Application Centric, possible. It also heralded the Windows® desktop as the presentation vehicle of choice. This marks the point in the evolution of technological process automation where the focus shifts from underlying technology to technology that facilitates the use of applications. Major advances in this decade were built on Internet technology. Probably the most significant is how Ethernet and TCP-IP enabled the concept of the "Common Information Infrastructure;" in essence this collapsed the traditional DCS architecture onto a common network backbone by allowing the process control and supervisory applications to interact in automation configurations in exactly the same way that Internet applications interact. It also provided a vehicle to integrate standards-based fieldbus technology into the same infrastructure and to exchange data using the same mechanisms.

Object Management, not to be confused with object programming, is a major beneficiary of this technology. It describes the ability to organize related data or information into a common structure and perform higher level functions on it and to interact with it as a defined entity; it is the cornerstone of CPAS and will be discussed in more detail later. Examples of the use of object management could be to group objects associated with a unit operation and perform global functions such as alarm suppression with a single action, or to create inheritance or association between different objects. It is an exceedingly important capability when the process to be controlled is functionally decomposed for advanced functions.

Until this decade there were an unlimited number of approaches to describing, designing, interfacing and programming applications. This decade delivered the common reference model for structuring process control applications—the ISA-88 series of standards, which is applicable to continuous, batch, and discrete process control. It also brought us a reference model for operations management applications (the ISA-95 series of standards), and finally it brought us an international standard, which organizes and prescribes the use of the most commonly used process control languages and configuration (IEC 61131-3).

Maintenance is the second largest controllable cost in a typical process plant and the fact that predictive maintenance has one tenth the cost of routine maintenance drove the need for *Plant Asset Management* (PAM) applications. These are essentially extensions of the control system and have been incorporated into most process automation systems with different levels of success. PAM began as a maintenance facility for automation field devices but now has a reach into automation assets as well as those assets that automation controls.

With the use of information technology in process control systems, it was inevitable that the human interface of choice (*Windows*) was not far behind. *OLE for Process Control* (OPC) technology also gave us a basis for interfacing Microsoft-based applications with both organized and disparate data and information. OPC has been recognized as a de facto standard communication protocol for non-mission-critical applications.

The Business Centric Decade

It could be argued that the previous four decades were a precursor to the Business Centric Decade, because at this point the focus changes from enabling technology to how to use available and proven technology to improve business performance. Other than as part of CPAS, discussion of these applications is outside of the scope of this chapter; however, the next section will discuss how the previously discussed evolution is being orchestrated into CPAS.

The Emergence of CPAS

When the benefits of utilizing information technologies in process plants became obvious, the major process automation companies all brought new systems to market, marking the evolution to the next generation and our current automation systems. Each differed, based on how a specific supplier viewed the problems to be solved and their perceived solution to those problems.

This led to a period of confusion for users, during which time several major operating companies retained the ARC Advisory Group to draft a vision for a system that supported Process Control, Advanced Process Control and Operations Management applications, complemented by human empowerment applications such as decision support and advanced analytics. It was required that this vision include general guiding principles and also provide enough detail to allow in-depth internal discussions as well as productive problem-solution discussions with external suppliers. This effort became the Collaborative Process Automation System, CPAS.

The *Guiding Principles* per se are outside the scope of this chapter but their essence is discussed here:

- The value proposition of the automation must be based on measurable improvements in the *Return on Assets* from the assets it controls.
- The automation must facilitate a *Continuous Improvement* culture.
- *Flawless Execution* of the automation strategy is the mandate.
- Everything that should be automated will be automated.
- The vision will deliver the lowest *Total Cost of Ownership*.
- A *Common Infrastructure* based on standards.
- No artificial *Barriers to Information* such as gateways or linking devices.

It was clear that the challenge was to deliver a capability that facilitated business performance. On the requirements side, these business factors included improving operational effectiveness, increasing asset utilization, minimizing variable costs, minimizing unscheduled downtime, preserving capital assets and reducing fixed costs. On the solution side, they equated to improving human effectiveness, empowering people and automation with predictive information, and enabling collaboration and synchronizing a real-time perspective in terms of a production cycle, a perspective that did not previously exist.

CPAS Views: Functional, Logical, Standard Application, and Advanced Applications

Functional View

Let us look at CPAS from a functional point of view. First, there are only two systems in a process plant, the business system and the automation system, each with different classes of applications. CPAS is an automation system; it is not hierarchical as was the Basic Process Control System (BPCS) of the past (Figure 1–13).

This is because the use of Ethernet TCP-IP allows it to collapse onto a single communications backbone with all applications, including the field devices, able to exchange data and information without barriers. This satisfies the CPAS principle of ***A common infrastructure, functionally transparent, logically concise and standards based,*** TCP-IP includes a protocol stack and manages the communications. ***This satisfies the CPAS principle of no artificial barriers to information.*** However, there are some good reasons for having some structure, such as consistency, deterministic operation and clarity. As has been said, this structure is provided for Process Control applications by the ISA88 Reference Model and for Operations Management applications by the ISA95 Reference Model. A third international standard is used for programming and configuration; this is IEC 61131-3, which organizes and prescribes the use of the five most commonly used process control languages. ***These facilitating standards contribute to the CPAS principle of flawless execution.***

There is also another dimension to data and information management across CPAS, and that is Global Data Access (GDA). Object Management (OM) was mentioned in the discussion of automation systems evolution; OM enables GDA. It has the following characteristicsIt requires every element of a CPAS to be addressable as an object with a unique name. This has the benefit

1.4 The ARC Advisory Group's Collaborative Process Automation System (CPAS) Vision

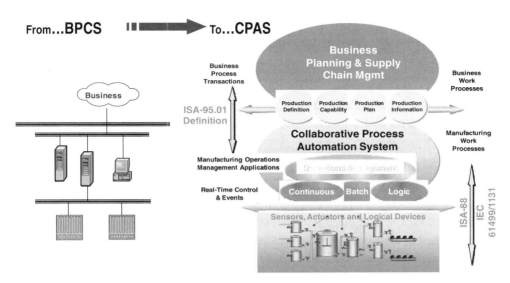

Figure 1–13 CPAS Functional View

of accessing data directly from the source and not requiring intermediate data stores. You always get the freshest data:

- Unlike tag-based systems, which require that paths between data intermediate sources (scanned intermediate data tables) and consumers of data be created during the configuration stage (early binding), CPAS utilizes late binding technology (dynamic linking) and publish/subscribe technology. This has the benefit of eliminating much of the configuration process because objects can be accessed using their unique descriptors at run time. It also makes the location of the source independent, which provides more flexibility in assembling and moving I/O components. A second benefit is that the system performs its own housekeeping when changes are made. In tag-based systems, the documentation cost can increase exponentially with the number of active system components.

- Since application technology is advancing along object lines, object-based systems provide more native environments for future applications.

- In tag-based control systems, control strategies are limited to the capacity of the controller they reside inside. Since object-based systems are hardware independent, they are very extensible, and extensibility only involves adding the underlying hardware. This cost usually correlates closely with incremental value.

- Data quality goes to the heart of system performance. CPAS uses quality tags to convey the value of data; if data has been corrupted, it is tagged for future operations and those operations are suspended, alarmed and/or alternate action is taken.

- In terms of hardware redundancy, CPAS takes it a step further, using Fault-Tolerant operation. This approach addresses the unique requirements of programmatic control because processors operate in lock-step and periodically marry. This means that if failure occurs in the primary processor, the secondary is never more than one operation behind. Also, because in systems that support Global Data Access there is no intermediate storage, all data exchanges are in parallel. This is the ultimate in software redundancy.

Logical View

We can discuss the CPAS architecture from a logical perspective (Figure 1–14). At the heart of this view is the Common Information Infrastructure (Figure 1–15), which has been discussed before as a single communications backbone. Since Ethernet TCP-IP has become the standard networking protocol for business, manufacturing and personal use, it can be called that single communications backbone. Obviously, security components provide separation but virtually, it is a common backbone.

Since this architecture is standards based it will have a long life cycle, and for the first time it will support an evolutionary approach to system upgrades and subsequent new generations of systems inevitably brought to the market by suppliers. In the past, these were replacement events.

This architecture is able to support a wide variety of functionality. For example, the advent of LAN (Local Area Network) based wireless will find this a native architecture to interface to. The standards based fieldbuses of FOUNDATION Fieldbus for process control and PROFIBUS and DeviceNet for logical control are easily interfaced through linking devices. Application-specific appliances such as analyzers, tracking devices and others can also be easily interfaced. Application servers supporting mission critical applications interface through redundant channels, and loosely coupled applications that are not mission critical are interfaced through OPC. And finally, since they share

1.4 The ARC Advisory Group's Collaborative Process Automation System (CPAS) Vision

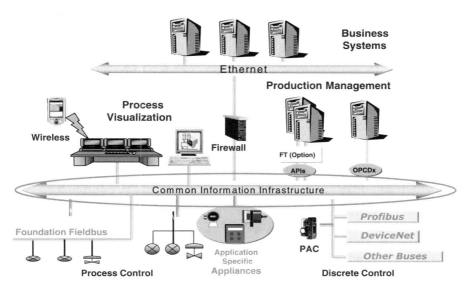

Figure 1–14 CPAS Logical View

- Common Logical Backbone (Message Bus)
  - Ethernet/TCP-IP
- Common Data/Information (Context)
  - Satisfies the 5 Any's
  - Relevance
  - Association (Dynamic Reference Model)
- Common Language (Semantic Content)
  - XML
- Common Presentation (Web Enabled)
  - Http
- Common Time (Event Synchronization)
  - Network Time (SNTP)
- Common Component Object (Interoperable Applications)
  - SOA or
  - OMG
- Unified Enterprise Work Processes (No Process Barriers)
  - Standardized Work Process Management

Figure 1–15 Components of the Common Information Infrastructure

the same logical backbone, Business Systems can communicate natively with Automation Systems.

Although CPAS is a single model with distributed processing, the configuration standard (IEC 61131-3) utilized in CPAS requires the use of common services, primarily System Management and Master Time. System Management is the facility that monitors the health of the system and reports any abnormalities. Master Time is the basis for all time stamps in the system; it is synchronized to the network time protocol in TCP-IP. Both of these shared services reside on a station in the system and there is a provision for each to be automatically reconstituted on a back-up station if the primary station fails.

Standard Application View

The application view is a framing of the applications at the Operations Management and Process Control areas. The Standard Application View of CPAS (Figure 1–16) displays some of the other principles. The application view is made up of applications that support the CPAS principles; they are divided into real-time functions and near-real-time functions. The organization of the real-time functionality is based on the control board having all of the manual responsibilities automated, which allows the operator to only become involved in the control of the process by exception.

Process Control and Monitoring

> **Progress and Performance Visualization**—The ability of the system to derive current performance and provide a single view of that performance, sometimes referred to as "a single version of the truth."
>
> **Troubleshooting**—Primary the responsibility of the field operator, with the board operator becoming involved as required.
>
> **Performance Monitoring**—Relates to Real-time Performance Management and is covered in a subsequent section.
>
> **Procedural Best Practices**—Relates to Procedure Automation. A high percentage of events and many unscheduled shut-downs are caused by a failure to follow prescribed procedures. The capability to effectively automate these procedures is a critical part of CPAS.
>
> **Operational Perspective**—The loss of experienced operators is a serious situation, and "best operators" are becoming fewer and fewer. Operational Perspective describes the capability of the system to relate current operation or possible changes in operation to a risk scenario derived with expert system technology. The objective is for all operators to perform as best operators.

1.4 The ARC Advisory Group's Collaborative Process Automation System (CPAS) Vision

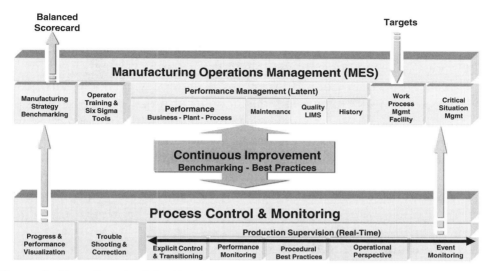

Figure 1–16   Standard Application View

Event Monitoring—CPAS manages events (both process alarms and equipment events) analytically, with the objective of relieving the operators of manual tasks and as a means to predict incidents that can be avoided or have their consequences mitigated.

Manufacturing Operations Management

Manufacturing Strategy Benchmarking—Most users have some form of a Balanced Scorecard to measure performance against targets; this feeds that application.

Operator Training—This function is a part of the user's Operational Excellence Program.

Performance Management—Performance management has changed considerably with the tools that are now available. This is discussed in a later section under Real-time Performance Management. This is also a functionality that CPAS is required to support.

Work Process Management Facility—CPAS has a fundamental principle to automate everything that should be automated; this applies to work processes as well as to manual tasks. This function is discussed in a subsequent section.

Critical Situation Management—The convention in the industry is to refer to events that lead to major economic losses, unscheduled shutdowns or HSE situations as incidents. These incidents have historically been almost impossible to predict and extremely hard for plants to recover from. However, by using newly available, low-cost wireless sensors to collect large amounts of process and asset data, combined with recently developed predictive analytic software to process the data, it is now practical to predict and preempt the majority of these incidents. This technology can, with a high level of certainty, predict a major incident either in time for an operator to avoid it or, if it cannot be avoided, to significantly mitigate the consequences.

CPAS Advanced Applications

There are several advanced applications that have emerged as critically important and are worth discussing.

Common View/Single Version of the Truth

If people are going to work effectively there is an increasing need for them to have accurate and timely information. And if their job involves collaborating with others then they need to be looking at the same information. This is referred to as "a single version of the truth" (Figure 1–17). The value of information is directly proportional to how fresh it is and how widely it is available to support collaboration. This level of information availability is not always the case; our research shows that in a typical process plant, only one person in twenty has the information they need to do their job. CPAS supports "a single version of the truth" used passively on thin clients to persistently on analytical workstations performing advanced decision support. Use of passive information usually relates to Ad-Hoc decision making; in this case relevant information is called up, a decision is made and then the information is released and not retained. Use of persistent information usually relates to analysis functions, performed on archived information; in this case the results are also retained in persistent storage.

Actionable Information

Automation systems routinely deliver too much data and too little information, and this is not helped by the Common Information Infrastructure. This is

1.4 The ARC Advisory Group's Collaborative Process Automation System (CPAS) Vision

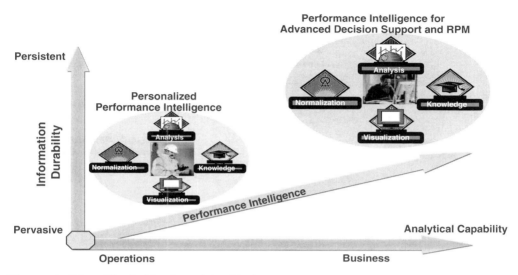

Figure 1–17   Single Version of the Truth

particularly true when the automation of work processes is required. In these cases, it is critical to obtain a perspective among several dimensions in order to make the correct decision.

Whether the work at hand is to automate a complex manual procedure or give the responsibility for this procedure to an inexperienced operator because of the absence of an experienced operator, the challenge is the same: "How can the procedure be executed correctly in a dynamic environment if the accurate context is not known?"

CPAS supports making the correct decision based on data accessed from the Data Model (sometimes Meta Data), the point you are now at in the work process, an accurate understanding of plant assets that are available, and finally the application that will be using the context, all with common time and assurance that good quality data is being processed (Figure 1–18).

Real-time Performance Management

In the past, performance was defined as producing a finite amount of a specific product in a particular time, with all targets set at the beginning of a shift. This is changing to meeting specific business targets for a specific product in a particular time, while monitoring plant performance in real time and continuously evaluating whether you can do better, or if there is a problem, how you

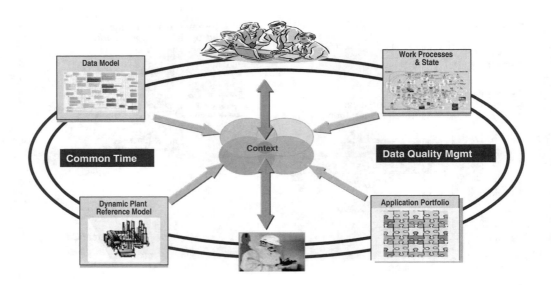

Everyone with a need to know should have visibility to information in context

Figure 1–18 Common Actionable Context

can recover by the end of the shift. This is referred to as ***"Real-time Performance Management"*** (RPM) (Figure 1–19) and is built on the concept of "Activity Based Costing" (ABC). ABC requires measurement of production rate and an understanding of the cost of production in order to have an understanding of the rate at which value is being added. This concept has been in use in discrete manufacturing for many years because all these values are known, but not in process manufacturing, because the value of intermediate products has not been known.

It has always been desirable to manage process plant operations in real time but providing the operators the tools has been a challenge. Now that real-time performance management has become critical, tools have become available to perform a functional decomposition of the process, correlate the process to the responsibilities of the operator, and provide an understanding of the operator's performance as well as the process's performance, all in real time. These tools include rigorous modeling, material balance, and optimization tools. CPAS is designed to support this technology.

1.4 The ARC Advisory Group's Collaborative Process Automation System (CPAS) Vision

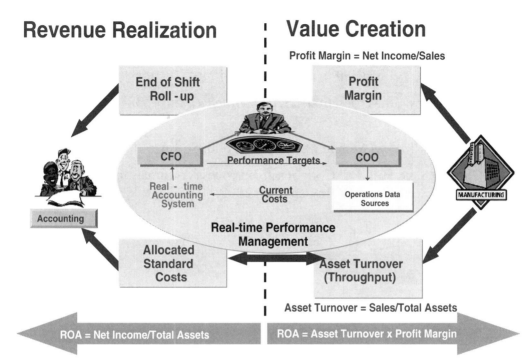

Figure 1–19   Real-time Performance Management

Where is CPAS Headed?

CPAS is currently a single model with distributed processing and shared services. It is based on the IEC 61131-3 configuration standard. It, along with all the major process control systems, is not a true distributed control system. It is distributed but not independent; all parts of the system are dependent on the shared services.

The next generation of CPAS will continue to support Process Control, Operations Management, and Advanced Decision Support, but it will do so as a single model with distributed processing separated into robust autonomous components. These components will be distributed and based on the new *IEC 61499 configuration and programming standard,* which does not require shared services, and it will be autonomous because it will support the new *Foundation for Intelligent Physical Agents (FIPA)* technology standard.

Bibliography

International Electrotechnical Commission, IEC 61131-3 *Programmable Controllers—Part 3: Programming Languages* Edition 2.0, 2003.

International Society of Automation, ISA-88 *Batch Control Parts 1–4,* 1995. See also IEC 61512.

ISA-95 *Enterprise-Control System Integration,* 2000. See also IEC 62264.

Woll, D., Caro, D., and Hill, D. "Collaborative Process Automation Systems of the Future." ARC, www.arcweb.com, 2002.

CHAPTER 2

2.1 CPAS System Architecture

Martin Hollender

Introduction

Process automation is a complex task requiring the smooth integration of a multitude of different systems and information sources. The following features and requirements are cornerstones for a CPAS architecture:

- **Integration**—Typical tasks like operations, engineering, and information management must all be available from one integrated environment. It must be possible to integrate external best-of-class tools.

- **Scalability and Flexibility**—To minimize system life-cycle costs, the design must allow for easy incremental addition of performance and functionality as needs evolve.

- **Investment Protection**—The architecture must ensure that future enhancements in systems technologies will not compromise the current investment and will provide users the ability to enhance their current system with future productivity-enhancing applications. The architecture needs to be open and standards-based (see Section 2.3).

- **Security**—Security is the condition of being protected against danger, with an emphasis on danger that originates from outside (see Section 2.7). Examples of security are protection against software viruses or terrorist attacks.

- **Safety**—Safety is the condition of being protected against danger, with an emphasis on danger that originates from inside (see Section 2.8). Examples of safety are protection against accidents or software bugs.

- **Traceability**—Traceability is the ability to reconstruct every production step (see Section 2.9). This is required both for regulatory compliance and as an important basis for production improvements.

2.1 CPAS System Architecture

- **Diagnostics**—Comprehensive system-wide self-diagnostics and reporting features result in ease of maintenance and maximum system availability.

- **Complete System Management**—Easy setup and maintenance ensure safe, reliable operations.

- **Dependability**—Dependability is the quality of a system to achieve a predefined behavior in the face of failures of its parts. It encompasses safety, availability, maintainability, and reliability (see Section 2.5).

- **Non-stop Availability**—In many plants a reduction or stop in production has significant associated costs. For example, petroleum refineries usually run for many years without interruption. This means that a CPAS needs to be designed in a way that upgrades and maintenance can be performed without disturbing production.

- **Globalization**—In today's globalized economy, a CPAS must be deployable everywhere in the world and must work with different regional settings such as languages, fonts, and time and date formats.

Object Orientation

One core problem for CPAS is the need to handle large numbers of items and their related information. These items include actuators, sensors and process equipment. For a larger plant, several hundred thousand such items may need to be organized. Starting in the 1960s, computer science has developed concepts for object orientation that are now standard in most modern programming languages and operating systems. Most of these concepts make sense for CPAS as well.

Previous generations of PAS have handled the items in long lists, and different applications were completely isolated from each other. As standardized by IEC 61346, objects relate plant data (the aspects), to specific plant assets (Figure 2–1). For example, aspects are associated informational items such as I/O definitions, engineering drawings, process graphics, reports, trends, etc. that are assigned to each object in the system. Aspect Object navigation presents the entire production facility in a consistent, easy-to-view fashion. This allows a single window environment to include smart field devices, plant asset management functions, information management, batch management, safety systems, and manufacturing execution system (MES) applications.

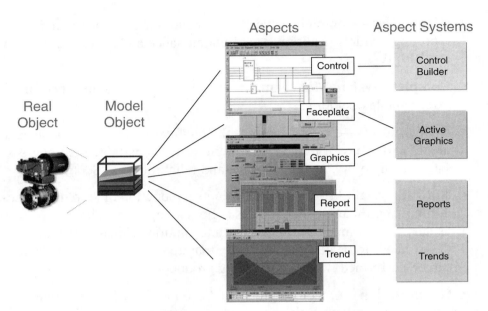

Figure 2–1 Mapping Plant Assets to Software Objects

Specific aspects like Plant Asset Management (see Section 5.4) can be added to a system later on demand. The Aspect Object Model can integrate data from external systems like enterprise resource planning (ERP) or computerized maintenance management systems (CMMS). Once such items have been mapped into the Aspect Object Model they are available for all other aspects. The Aspect Object Model can serve as a basis for the integration of different applications:

1. Let's assume there is an object representing a pump with an asset monitoring aspect; the pump needs the date of the last maintenance as the basis of a diagnostic algorithm.

2. This aspect now tries to find a CMMS aspect on the same object.

3. If it is found, the asset monitor connects to the CMMS aspect and asks for the date of the last maintenance.

4. The asset monitor then runs the diagnostic algorithm.

5. If a problem is found the asset monitor aspect automatically opens a maintenance ticket on the CMMS aspect.

2.1 CPAS System Architecture

This example demonstrates how more-generic solutions can be implemented. Of course it would be possible to directly access the CMMS system, but such a solution would be very project specific and not deployable as a general solution. The abstract information access possible with the Aspect Object Model allows a modular integration of applications. Object orientation is explained in more detail in Section 2.2.

Object Types and Object Instances

The Object Type concept (Object Types can also be called classes, templates, or typicals) allows the preparation of abstract solutions that can be instantiated several times or even be reused in other projects. An Object Type can contain connectivity to plant devices, faceplates, alarm lists, asset monitors, documentation, custom script code, and many more. Object Types allow packaging proven best-in-class solutions and deploying them to many other sites and projects where the same solution is needed. Object Instances are instantiated Object Types and correspond to existing real-world objects, whereas an Object Type is a more abstract concept.

A newly created object instance (Figure 2–2) that was derived from an Object Type immediately has all data and functionality that was prepared for the Object Type. Required changes like bug fixes or modifications need to be implemented only once on the Object Type and are then automatically propagated to all instances.

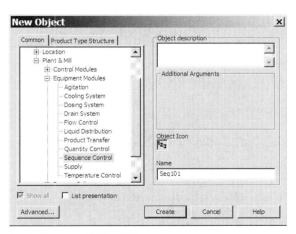

Figure 2–2 Instantiating an Object Type

It is possible to stepwise refine Object Types. For example, an Object Type for a proportional-integral-derivative (PID) controller can be customized and extended for the needs of specific industries. Project-specific aspects can easily be added to object instances.

The Object Type concept helps to implement standardized solutions, because the object instances inherit their methods and properties from the Object Type. Applications can therefore assume the availability of those standardized properties and methods, which allows the creation of more abstract and generic applications. With help of the Object Type concept, it is possible to find all instances of a certain type. For example, all valves of a certain type can be identified. This kind of functionality is not only useful for operators during daily plant operation, but also for applications that can be written to be more generic if they can use Object Type information to identify relevant objects.

Service Framework

A main task of a CPAS is to provide reliable access to a set of distributed services. Popular operating systems like Microsoft Windows or Linux do not provide a complete high-quality, real-time, secure and robust platform for distributed computing as needed by a CPAS. Originally the hope was that Microsoft DCOM™ technology could be used, but it turned out that it had several severe problems. CPAS vendors were forced to either replace DCOM or to modify it to make it suitable for the requirements of CPAS.

Important mainstream applications like server farms for huge web shops or database servers have slightly different requirements than CPAS. Standard operating systems such as Microsoft Windows can be used as a basis for CPAS servers and workstations but need to be extended with a framework suited to the specific needs of process automation, which might include the following:

- Monitor the health of all services
- Distribute data and function calls between different nodes of the system
- Manage redundant services
- Distribute load between redundant servers
- Restart and re-initialize failed services

2.1 CPAS System Architecture

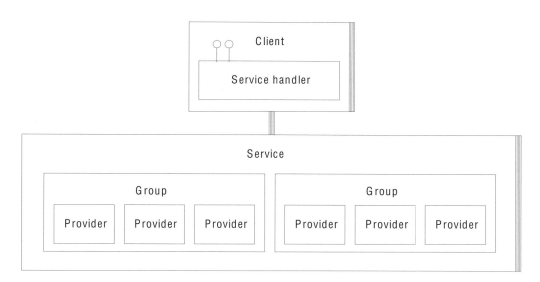

Figure 2–3   CPAS Service Framework

- In case of errors, clients need to be transparently switched-over to redundant services

A provider is a specific service running on a server node. Several redundant providers are pooled in one service group (Figure 2–3). If a service provider fails, the framework automatically reconnects clients to a working provider in the same group. Once the provider is working again, clients can be reconnected with the original service provider.

Connectivity to Automation Controllers

For connection with automation controllers, the OPC (open connectivity) Data Access (DA) and Alarm & Event (AE) standards (see Section 2.3) are well-established industry standards supported by a large number of vendors. In larger plants, many different automation controllers, often from different vendors and different technology generations, need to be integrated (e.g., power plants often consist of several units built at different times). It is an important task of the CPAS to harmonize the associated OPC servers under one umbrella. With the help of an umbrella OPC server, applications don't need to bother on which OPC server a certain tag can be found. (Without an umbrella framework, an application would need to first address a specific server before it can address an OPC tag. Applications with hardcoded server addresses are less robust against change and less portable.) The CPAS umbrella OPC server consolidates all underlying OPC servers (Figure 2–4). The umbrella server is available on all nodes of the system. It is the only OPC server a client needs to talk to.

Access to data from different sources can be unified. The client application only needs to know the tag name and doesn't need to use DCOM to access the OPC server. The framework knows on which OPC server the tag is residing and handles all communication with the OPC server. The key tasks of the consolidating umbrella framework are:

- Registration of the items available in an OPC server in a central repository, so that a search for a tag doesn't require the scanning of all connected OPC servers

- Automatic enrichment of OPC tags with value-adding features like faceplates, alarm lists, or asset monitors.

The plug-and-play connectivity concept allows many different OPC servers to be placed under one common server (Figure 2–5). If an automation controller doesn't have an OPC server, the preferred way for integration is to create an OPC server for this controller.

Object Lookup and Access Service

As a basis for generic and portable solutions, it is important that the access paths to information are not hardcoded into applications. Mechanisms are needed that allow a general lookup of information independent from where this information is located. The most basic such service is to retrieve objects by name. But more complex services also need to be available, such as identifying all control loops of a certain type. Queries should be able to include structural information. For example, it should be possible to retrieve all motors in a certain plant area or retrieve the controller that a specific I/O point is assigned to.

Objects need to offer generic mechanisms that allow accessing information from the object without knowledge about the application or where the information is coming from (Figure 2–6).

2.1 CPAS System Architecture

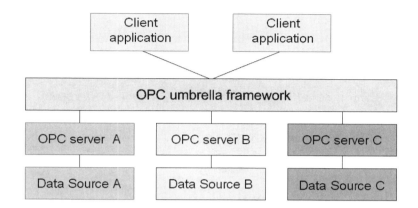

Figure 2–4   Consolidating Umbrella OPC Server

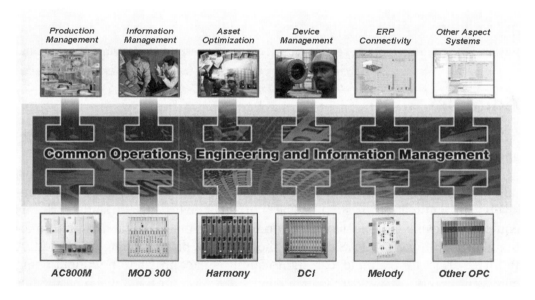

Figure 2–5   CPAS Plug-and-Play Connectivity

This abstraction allows the building of applications that are more generic. For example, it should be possible to retrieve a real-time value without knowing which automation controller is actually used, and it should be possible to retrieve the last maintenance date of a pump without knowing if the CMMS comes from SAP, IFS, or IBM.

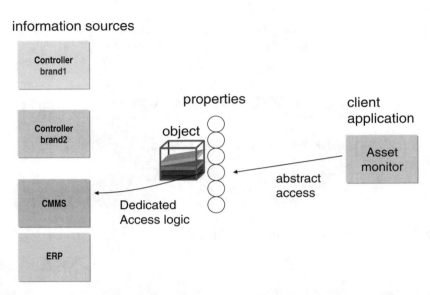

Figure 2–6   Abstract Properties Allow Generic Access to Underlying Applications

Administration

Administration includes startup and shutdown of the CPAS, health monitoring, and fault diagnosis.

Figure 2–7 shows a display with health status information for all important system services. Standard computer and network equipment is used extensively in automation systems. The correct behavior and status of this equipment have a significant impact on the performance, reliability, and functional availability of a CPAS and thus the industrial process being controlled. A full hard drive, for example, can significantly impact production without being evident. Monitoring software can independently monitor the status of computer and network equipment. Simple Network Management Protocol (SNMP) is the standard communications protocol used to monitor and manage network devices. If problems are discovered, it can be announced to the administrators via an alarm, email, or shore message service (SMS).

Users and their access rights need to be administrated. This is explained in more detail in Section 2.7.

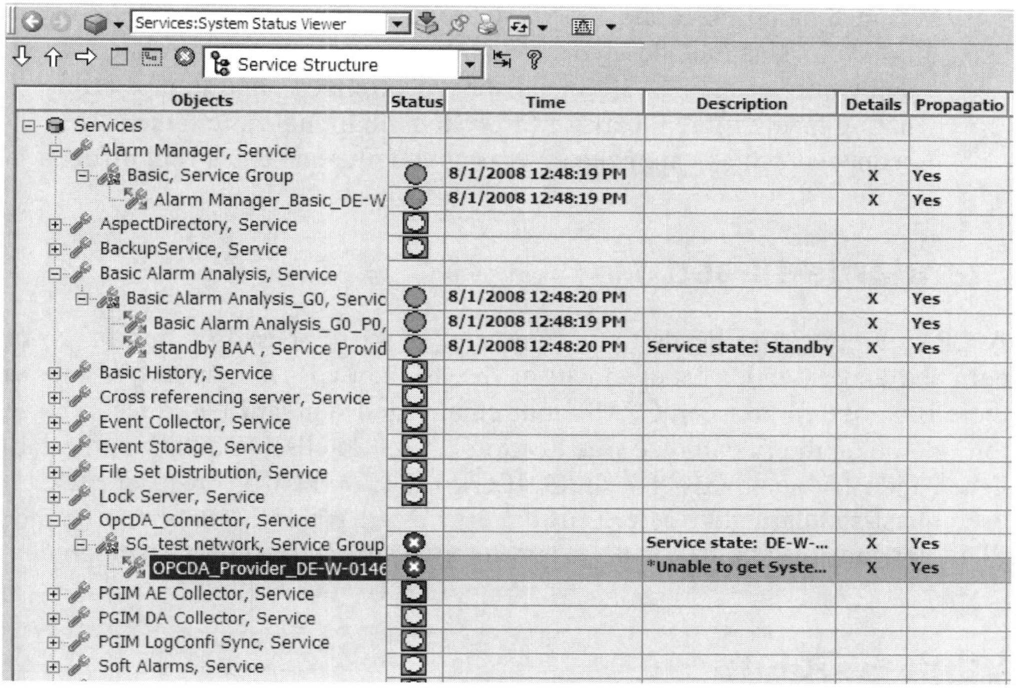

2.1 CPAS System Architecture   51

Figure 2-7   Health Monitoring of System Services

Basic Services

The service framework hosts all services required around the clock. Important standard basic services include:

- Object Services: Run the intelligence in the system including services related to object management, name searching and browsing, and security.

- Connectivity Services: Provide access to controllers and other data sources. Examples of services that run on a connectivity server are OPC-related services (DA, AE, and HDA—Historical Data Access) and system messages.

- Application Services: Run various types of system applications (e.g., batch management, asset optimization, simulation and optimization, and information management). Running application services on dedicated servers minimizes interference with the core system.

- Time Synchronization entails synchronizing the time on the nodes in the network. This includes synchronizing the automation controller, server, and workplace nodes. If time is not synchronized, time stamps originating from different parts of the system might not fit into the correct sequence, which can make retrospective failure analysis very difficult.

Standards-based

A CPAS architecture needs to be open and standards-based so that it can be enriched and extended independent of the original CPAS supplier. Whenever there is a good fit between CPAS requirements and standard-based technology, this should be the first choice (see Section 2.3 for details). In addition to established industry standards like OPC, IEC 61131, or ISA-95 there are widely used quasi-standards like Microsoft Excel™, MathWorks Matlab™, or Adobe PDF™, which also help to make a system open and fit for future development.

Logical Colors

Consistent use of color is a key issue for process operators. Colors need to be configurable because different plants, companies, industries and countries want to use different color schemes. If the CPAS should allow direct assignment of RGB (red-green-blue) color codes, this would not be a good idea, because all items where each color is used would need to be changed, which might mean a lot of repetitive work. Using the color palette of the operating system is also not an option, because these settings are valid for a single workplace only. The CPAS can solve this problem by providing centralized "logical colors." Logical colors map names (e.g., alarm_color1) to RGB values (e.g., 255-0-0). Applications then specify the name of the logical color. If later on the RGB value for a logical color is changed, all display elements on all workstations using the logical color automatically update to the new color.

Software Globalization

Globalization has been one of the most important trends during the last decades. Many multinational corporations now have production plants all over

2.1 CPAS System Architecture

the world. Most CPAS vendors sell their systems worldwide. In software engineering, the methods and techniques used to make software adaptable to different languages and cultures are called globalization (e.g., IBM globalization and Microsoft Go Global) or internationalization, and localization or Native Language Support (NLS). Internationalization covers more generic coding and design issues, whereas localization involves translating and customizing a solution for a specific market.

The use of English as a lingua franca can solve some CPAS globalization issues, but not all. For example, many operators prefer to use their native language and some operators don't speak English at all.

The major issues with CPAS globalization include:

- Support for international fonts like Japanese, Arabic and Swedish
- Text in the CPAS needs to be translatable
- Date/Time formats need to be adaptable (Figure 2–8)
- Different cultures use different separators: comma or semicolon

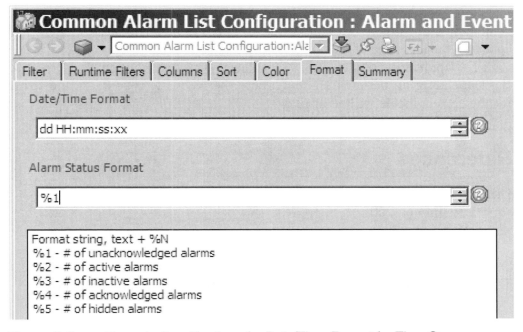

Figure 2–8   Example Specification of a Date/Time Format for Time Stamps

Many globalization issues are relevant not only for CPAS software, but for software in general. One very important principle is to code all strings as Unicode resource strings. Unicode allows the representing of text in most of the world's languages, including Chinese and Arabic. Having strings in separate resource files allows efficient translation into other languages. The resulting software package can then be shipped in versions for the different languages.

NLS Resources

In a CPAS software application, objects like libraries, displays, faceplates, or display elements often contain strings presented to operators. In today's globalized production, it is very important that solutions containing valuable best practice know-how can be deployed to production plants all over the world.

It is not manageable to create language-specific versions of these application objects, because this would lead to an explosion of the number of software packages, which could not be maintained efficiently. NLS resources (Figure 2–9) allow strings to be stored in several different languages, with only the string matching the target environment being displayed. Only one software package containing strings in all relevant languages needs to be maintained; the text will appear in the language of the operators.

The application object (e.g., a faceplate) refers to a string via the Resource ID. Depending on the environment where the faceplate is running, the string for the correct locale will be displayed.

References

Dr. International *Developing International Software*, Microsoft Press: Redmond, 2002.

IBM Globalization www.ibm.com/software/globalization (last retrieved 22.05.2009).

Microsoft Go Global Developer Center msdn.microsoft.com/goglobal (last retrieved 22.05.2009).

2.1 CPAS System Architecture

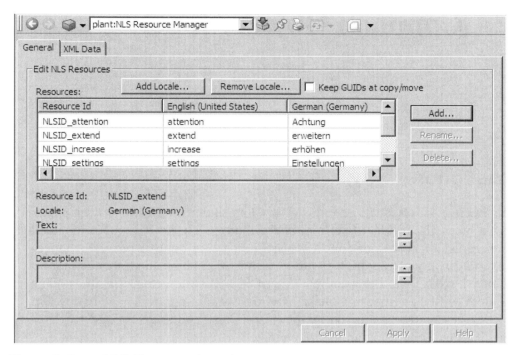

Figure 2–9   NLS Resource Aspect

Bibliography

International Electrotechnical Commission, IEC 61346, *Industrial Systems, Installations and Equipment and Industrial Products—Structuring Principles and Reference Designations* 1996.

Samad, T., McLaughlin, P., and Lu, J. "System Architecture for Process Automation: Review and Trends." *Journal of Process Control 17*, 2007: 191–201.

2.2 Common Object Model

Martin Hollender

Introduction

As has been mentioned, large industrial plants like refineries or power plants can have many thousands of sensors and actuators. These physical items need to be mapped to corresponding software items like control programs, faceplates, and display elements. It is obvious that such large quantities cannot be efficiently handled without systematic naming and structuring principles. This is true both for the engineering phase, where several integrators need to develop a common understanding of the plant (Garcia and Gelle, 2006) and for later phases like plant operation and maintenance.

Structuring is a way of managing complexity. IEC 61346 uses the following important concepts for structuring: object, aspect, and structure. Structuring is defined as a way to organize the objects of a system in a systematic way, in order to facilitate all the activities that need to be performed during the entire life cycle of that system. With structuring as a tool, it is possible to define reusable objects—software engineering solutions—which, especially if computer-aided tools are used, can easily be used as building blocks. This in turn provides the possibility for feedback on solution experience, and thus to increased quality.

IEC 61346 defines an object as an entity treated in the process of design, engineering, realization, operation, maintenance and demolition. An aspect is defined as a specific way of selecting information on or describing a system or an object of a system (IEC 61346-1996). A reference designation must unambiguously identify an object of interest within the considered system.

The objects can be put in hierarchical structures (Figure 2–10). The branches represent the subdivision of these objects into other objects (i.e., sub-objects). Each object that occurs within another object is assigned a single-level reference designation unique with respect to the object in which it occurs. The object represented by the top node is not assigned a single-level reference designation.

2.2 *Common Object Model* 57

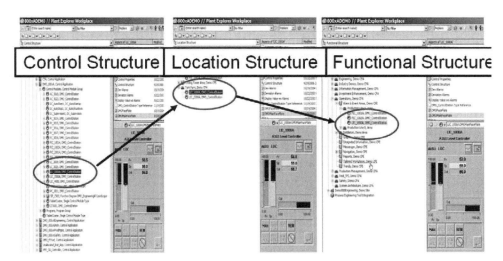

Figure 2–10   The Same Object Viewed in Different Structures

IEC 61346 establishes three general-purpose fundamental structures that can be applied to any technical system. These are as follows:

- **Function-Oriented**—Structure where objects are organized according to their functional aspect. It answers the question of what each object does, and it is identified by the equals sign.

- **Location-Oriented**—Structure where objects are organized according to their location aspect. It answers the question of where each object is located, and it is identified by the plus sign.

- **Product-Oriented**—Structure where objects are organized according to their product aspect. It answers the question of how each object is constructed, and it is identified by the minus sign.

Figure 2–10 shows an example of how the principles defined in IEC 61346 can be used to structure information in a CPAS (the "control" structure shown is an extension to the standard).

Object orientation attempts to address system complexity by emphasizing modularity and re-usability. Software objects are self-sufficient modules modeling all the information needed to manipulate the mapped real-world object. Software objects allow users to view any physical component of the automation scheme, from a pump or valve to a process unit or pressure transmitter, as a software object. Each object in the system has a number of attached aspects that can range from integration to computerized maintenance management systems (CMMS) and enterprise asset management (EAM) systems to schematic drawings and trending information. Objects provide the key real-time linkage between equipment and applications.

Object technology associates information with the plant and business entities (the objects) it belongs to, by organizing these objects to mirror reality and by providing powerful functions for navigation and searching. An electric motor, for example, can be represented by an object containing all the information connected with it. This might include design drawings, control diagrams, maintenance information, location, quality information, configuration information and measured values. It is important to realize that an aspect is not just the information connected with a particular Aspect Object; it also defines a set of software functions that create, access, and manipulate this information.

The **Common Object Model** addresses the issue of presenting information and allowing a user to operate on information in a consistent way. The model also addresses how different functions are integrated into the system in a way natural to the user. The Common Object Model is based on Aspect Objects and Aspects.

The real-world objects in the model are the objects the user interacts with. For example, such an object can be a reactor, a pump, or a node (computer). The software object is a container that holds different parts, aspects, of the real-world object. An aspect is one 'piece' of data and operations that is associated with an object. Typical aspects of an object are its control program, operator faceplate, trend configuration, function specification, and so on.

All interactions with an object in the system are carried out with the help of its aspects. Aspect Objects allow us to model a user's world in a way that is natural to him or her. We do this by building relations between objects (most often hierarchical relationships) and associate the necessary aspects with the different objects. Software components can be integrated in a seamless way with the help of the loose association between objects and aspects but also through the way new types of aspects are brought into the system.

Benefits of a Common Object Model for CPAS

In the past, with monolithic PAS solutions from a single vendor and a focus on Levels 0-2 in the ISA-95 model, most PAS have worked with a flat list of tags. Integration with other systems was rarely a requirement and if it became one, the integration could be solved with (often very expensive) custom programming.

2.2 Common Object Model

Since then the requirements have dramatically changed. Many plants use several different automation controller families from different generations and different vendors. A highly optimized production requires that all relevant systems are tightly integrated. This means that a CPAS needs to offer a Common Object Model infrastructure as a basis for integration of a heterogeneous multi-vendor system landscape.

Unified Access to Heterogeneous Data

Relevant data is scattered around in MES, CMMS, LIMS (Laboratory Information Management Systems) and ERP systems. Without a Common Object Model, each of these systems would need to be accessed individually. Direct access to the underlying systems also means that the server where the data resides must be explicitly specified. This reduces the flexibility and scalability of the solution, because servers cannot easily be changed. Solutions are difficult to transfer to new projects. The CPAS Common Object Model acts as a layer of abstraction above heterogeneous underlying systems. Applications working against the Common Object Model can focus on their core logic and don't need to bother about how to access the underlying systems. If they would connect directly to the underlying systems, the applications would need to contain a larger portion of access logic, making the applications less transferable to other contexts, where different underlying systems might be used.

The Common Object Model allows the system to address objects in a unified way, no matter where they are physically located. Data can be accessed via the Common Object Model so that applications can be written to be more abstract and reusable in different context. The name service of the Common Object Model resolves names from objects in a way that is independent of the server where the object actually runs. The unified application programming interface (API) shown in Figure 2–11 is based on the Common Object Model and can be used both by core functions of the CPAS itself, such as graphic displays and by project-specific extensions (e.g., a connection to a legacy system).

Object-Oriented User Interaction

Object-oriented user interaction with direct manipulation became popular with the introduction of the Apple Macintosh and later with the growing success of Microsoft Windows. These operating systems allow the user to point to an

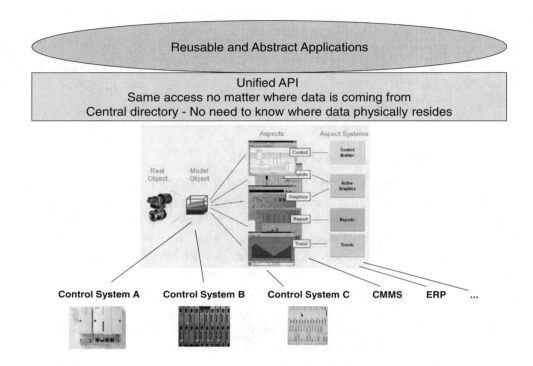

Figure 2–11 The Common Object Model Can Harmonize Access to Different Systems

object of interest (e.g., a file) and directly start all operations of interest in the context of that object (e.g., to copy, rename, or delete the file). Younger computer users take object-oriented interaction for granted.

Today users also expect the same direct interaction in a CPAS. For example, they want to be able to click on an alarm line and directly access related functionality in the context of that alarm (e.g., open a faceplate or read troubleshooting documentation) (Figure 2–12). If a CMMS has been added to the system, access to it should be directly available in the context menu. If an alarm indicates that a pump is broken, the context menu in the alarm list should allow the operator to open the CMMS directly for the pump in question and to conveniently issue a maintenance ticket.

2.2 Common Object Model

Figure 2–12   Object-Oriented Context Menu in a CPAS

Enabling Generic Solutions

An algorithm used to monitor how well a control loop of a certain type is performing might only require a few values like set-point, deviation and current output. The Common Object Model enables the system (or the operator) to find all loops of the type in question and then to access the required variables. This works independently of the hardware on which the loop is running. More complex scenarios include access to a CMMS to find out when a device was last serviced or to identify all devices with a specific serial number. This only works on the basis of a rigid and consistent standardization. For example, the date of the last service needs to be accessible at exactly one well-defined place. Given the high complexity of automation solutions, thousands of such standardizations need to be defined and enforced. More techniques to support generic and reusable solutions are discussed in the following section on object orientation.

Object Orientation

In computer science, object orientation has been an important paradigm for many years. Object-oriented programming languages like Smalltalk, C++, and Java™ support the construction of object-oriented software systems. Object orientation closely reflects the structure of the real world and is well suited to model complex systems with complex behaviors. Object orientation in the context of control systems engineering will be discussed in Section 3.3. Good object-oriented design is not easy and requires a lot of standardization between the involved parties. In the following, we discuss how the principles of object orientation can be used in a CPAS to create better solutions. The CPAS provides object-oriented base mechanisms so that object-oriented solutions (e.g., an application library for the Oil and Gas industry) can be built without any C++ or Java programming.

Type/Instance

The class/instance concept of object-oriented programming is nicely reflected in object types and instances. All instances of an object behave as defined in the object type, and if the type is changed, all related objects change with it.

In addition to just implementing the functional behavior of an object, other aspects of that object can be associated with the object type, such as operator station functionality (faceplates, etc.), documentation, alarm lists, trend displays, and so on. A well-defined object type therefore reduces engineering time significantly, because all objects instantiated from such a type definition are automatically equipped with the pre-defined aspects. If this is done consistently throughout the implementation, the operator benefits from a common look-and-feel, and easily finds the same information associated with every object. This also adds to operator efficiency.

To consistently maintain sets of object classes that relate to each other (device objects, common functions, and standard automation functions), these classes are typically maintained in libraries. These libraries are then imported into a project to start engineering on a relatively high level of abstraction.

Objects typically available in libraries are:

- Device control functions, such as motors, pumps, valves, fans, and so on.
- Process control blocks (e.g., proportional-integral-derivative [PID] loop and set-point control).

- Frequently used mathematical functions (e.g., steam pressure calculation, valve non-linearity compensation, and so on).

But complete process functions can also be maintained in libraries to allow an even higher level of re-use in customer projects.

Maintaining libraries independent of a customer project does require some up-front effort. But if libraries are tested and maintained as such and not related to a project, this may greatly add to the quality of the implementation. If specific test applications are designed to fully test library objects in a test-only framework that calls on all use cases, the test coverage is wider than what is possible in a project framework. Furthermore, automatic testing could be introduced for widely used standard functions to further reduce maintenance efforts.

Inheritance

The concept of inheritance was developed for object-oriented programming languages like Eiffel (Meyer, 2000) or Smalltalk. Today, it is a standard part of most modern programming languages. The idea of inheritance allows programmers to build generic super classes describing those features and properties that several other more specialized classes have in common. Such a more specialized class can then inherit the generic features and properties from the super class(es).

In Figure 2–13, the two classes "motor" and "valve" are derived from the super class "device." The classes "AC motor" and "DC motor" derive from "motor." "DC motor 123" is an instance of the class "DC motor" representing a real existing DC motor. It inherits the algorithm from the class "DC motor" and the property "RPM" from the class "motor." The properties "manufacturer" and "serial number" come from the inherited super class "device" (polymorphism). Sometimes the content of an inherited property should be identical for all instances (e.g., the manufacturer of a specific AC motor type is the same for all instances of that motor type. In this case both the definition "manufacturer" and its content "manufacturer-xy" are inherited to the instances. It is impossible to change the property on instance level.) For other properties only the definition of the property is inherited, whereas the content of the property needs to be set on each instance. This is the case for example for the property "serial number" because each instance will have a different serial number. The value of the super class defining the property "serial number" is that all instances consistently will have such a property.

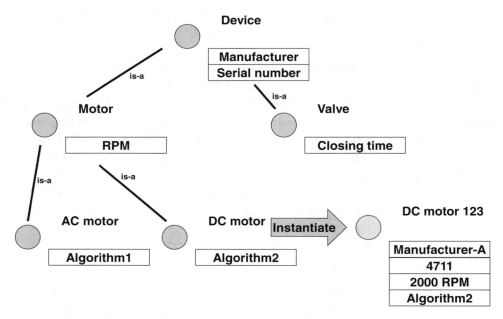

Figure 2–13 Class Inheritance Diagram

Inheritance allows implementing generic functionality. For example, an algorithm can use the property "serial number" for all instances directly or indirectly derived from the class "device." Generic features and properties should be positioned as high as possible in the class hierarchy so that algorithms using those features and properties can span as many objects as possible.

Object Types in a CPAS can include aspects from several applications and package abstract best-practice solutions. These Object Types can then be loaded into a project-specific environment and objects can be instantiated from the Object Types. Many of the aspects of the Object Type can be inherited as they are, whereas some others may need project specific configuration.

Encapsulation

The principle of encapsulation is to hide internal details from outside entities. Interaction with encapsulated modules is only possible via well-defined interfaces. Encapsulation increases the robustness of a system, because changes should not unexpectedly influence distant system parts; the only way the system parts can interact is via the well-defined interface. This interface should be designed abstract and thin, meaning the data structures should be logical and contain only the minimum of required data.

Encapsulation also supports testability, because tests can focus on the functionality provided via the interfaces. Encapsulation makes a system more agile, because the internal implementation of a module can be changed if the interfaces stay constant. Encapsulation forces solution providers to implement critical functionality inside the borders of a module. The analysis of the functionality can focus on the module and the interfaces. It is impossible that an important aspect of the critical functionality will be implemented somewhere else in the system, because the internals of the module are inaccessible to other system parts. Designing good encapsulation with well-defined interfaces that precisely expose all relevant functionality, but not more, is rather difficult.

As an example of encapsulation, let's assume that the main task of some module is to implement a sorting algorithm. Without encapsulation, the module would expose its internal data structures, and external clients might directly access those data structures. If later on an algorithm with better performance needs to be implemented, it is not easily possible to change the data structures, because external clients might rely on them. If encapsulation has been used, the results can only be accessed via a well-defined interface and therefore the internal data structures can be changed without much risk.

Modularity

Modularity is a concept coming from systems engineering. Large systems are decomposed into smaller independent sub-systems. Modularity refers to the tightness of coupling between components. The decomposed modules are easier to understand and to manage than the complex overall system. Modularity is closely related to encapsulation. Modules can be easily replaced and can be used in different context. For example, let's assume that a project has created a functionally excellent solution to solve a specific problem. Let's assume further that the solution is not modular. This means that the solution heavily relies on the specific environment of this project. Despite the fact that it is a solution with excellent functionality, it is very difficult to transport the solution to other projects, which would be interested in the functionality, but have different environments.

Polymorphism

Polymorphism, which was mentioned earlier, is another important concept of object orientation. It means that instances of different classes can be treated the same way, for example that they display a similar subset of properties. Usually this is enabled by both classes deriving from the same superclass (see Inheritance above). In the example above, even if the motors are of different types, they share the same property for serial number. Polymorphism allows the implementing of more general applications.

References

Garcia, R.G. and Gelle, E. "Applying and Adapting the IEC 61346 Standard to Industrial Automation Applications" *IEEE Transaction on Industrial Informatics* 2006: 185–191.

Meyer, B. "Object-Oriented Software Construction" Upper Saddle River, NJ: Prentice Hall, 2000.

International Electrotechnical Commission, IEC 61346 *Industrial Systems, Installations and Equipment, and Industrial Products—Structuring Principle and Reference Designations. Parts 1–4*, 1996.

2.3 Industry Standards and Open Interfaces

Martin Hollender

Introduction

Historically, process automation systems (PAS) were monolithic applications based on proprietary technology. As the technology matured, it became clear that the use of standards and open interfaces offers important advantages:

- Reduced total cost of ownership.

    o   Protection of investment. Once a proprietary PAS is discontinued or no longer supported, it becomes more and more difficult to extend and maintain existing installations.

    o   Reduced training cost. If personnel have been trained to master a standard, they can work with equipment from different vendors adhering to that standard. In the case of proprietary equipment, a basic retraining for every vendor is necessary.

- Architectural flexibility. A monolithic PAS offers only the possibilities foreseen by the PAS vendor. Everything else requires either very expensive custom developments or is impossible.

- Evolutionary enhancements. Industrial production facilities change over time, control strategies change, new knowledge evolves and new tools become available. One example is the replacement of the now-obsolete hardware of operator stations. The repair cost of these 15- to 20-year old stations is so high that the operator stations need to be replaced with current generation PCs. A monolithic PAS requires a replacement of the whole system, including the controller layer, even if this layer is still working fine. Standards and open interfaces allow adapting a CPAS as evolutionary enhancements become desirable.

- Interoperability. Today's production facilities usually still have lots of information islands, where different systems are not able to easily exchange information, although this exchange would be very beneficial for production.

- Mixed suppliers. If a production company standardizes on proprietary technology from a single vendor, this means once the decision is made, it can only be revised with enormous cost, because most of the automation equipment will need to be replaced. This vendor lock-in will probably result in higher prices and decreased quality, because the automation vendor knows that the buyer has no other choice. If sufficient standards are in place, this allows production companies to combine best-in-class components from different automation vendors.

- Interchangeability. If standard components from different vendors can be used, the resulting competition leads to higher quality and less cost. If a component is no longer being produced, it can be replaced by a different component from another vendor.

As these advantages are significant, modern CPAS are heavily based on standards. Standards promote choices, increase performance, and reduce risk. A huge number of standards exist, both in process automation and information technology. It is beyond the scope of this book to describe all relevant standards. The aim of this chapter is to introduce some of the most important standards relevant to CPAS. For those students in the area of automatic control who have insufficient knowledge about information technology, standards like OPC and XML will be explained in a bit more detail.

General Standards with High Relevance for Process Automation

In the following sections, several standards are briefly described that were originally created for areas other than process automation, but have proven to be very useful in the context of process automation. These standards are now part of the tool box of today's automation engineers. Later in the chapter, more specific process automation standards will follow. As several of those specific standards are based on general standards, it makes sense to introduce the general standards first, even if the importance of the specific standards for process automation is higher.

XML

The Extensible Markup Language (XML) is a simple and flexible text format playing an increasingly important role in the exchange of a wide variety of data. The W3C (World Wide Web Consortium) created, developed, and maintains the XML specification (w3c.org/xml). The W3C is also the primary center for developing other cross-industry specifications that are based on XML.

XML is used more and more in the context of process automation. Prominent examples are:

- ISA-95 (see below)
- OPC UA (see Section 2.4)
- AutomationML (see www.automationml.org)

XML offers a highly standardized approach for storing structured data and has excellent support of all kinds of tools. Modern database systems include more and more support for XML. A good overview on XML and related standards can be found in Skonnard and Gudgin, 2001.

XML Schema

An XML schema can be used to express a set of rules to which an XML document must conform in order to be considered "valid" according to that schema. The rules provide a means for defining the structure, content and semantics of XML documents. A popular and expressive XML schema language, XML Schema ("Schema" must be capitalized when this language is referred to), is maintained by the W3C (www.w3c.org/XML/Schema). An XML Schema instance is an XML Schema Definition (XSD) and typically has the filename extension ".xsd." For example: the Organization for Production Technology, formerly the World Batch Forum (www.wbf.org) has published the Business To Manufacturing Markup Language (B2MML) schemas, based on XML Schema. Support to validate XML documents against schemas can be implemented by tools such as XML editors or browsers.

XSLT

The Extensible Stylesheet Language Transformation (XSLT) is an XML-based language used for the transformation of XML documents into other XML or "human-readable" documents (Figure 2–14). For example, an XSLT transformation can be used to convert data between different XML schemas.

Microsoft Windows has an XSLT processor built in, and many other implementations, such as Apache Xalan, exist.

In the example shown in Figure 2–15, a XML document containing tag names, values, and thresholds is transformed into a comma separated list. The XSL style sheet controls how each XML element is transformed into the result document. Often simple style sheets, like the one shown in the figure, can be used to solve real-world problems, but also much more complex transformations exist.

Binary, CSV, and XML Data Exchange

The transfer of process data between different applications is one of the basic tasks in process automation. Of course, standards like OPC (Open Connectivity) support such tasks, but often the data needs to be transferred in files.

There are two character-encoding schemes in use: ASCII and Unicode. The American Standard Code for Information Interchange (ASCII) defines a 7-bit code for 128 characters from the English alphabet. It is still the most popular character-encoding scheme for raw text. The problem with ASCII is that it can't represent non-English characters like Chinese or Arabic, or German umlauts. Unicode (The Unicode Consortium 2006; www.unicode.org) provides a unique number for every character, no matter what the platform, no matter what the program, no matter what the language.

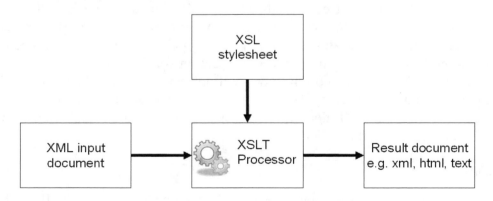

Figure 2–14 XSLT Transformation

2.3 Industry Standards and Open Interfaces 71

Figure 2–15   Example of How XSLT Can Be Used to Transform an XML Document

Unicode has been widely adopted by the computing industry and is the official way to implement ISO/IEC 10646. The file size of raw text encoded in Unicode is larger than that of text encoded in ASCII (usually twice as large) but in most cases the gained generality outweighs the increased space consumption. Program routines to read text need special considerations to be able to read Unicode text, but in today's globalized economy, Unicode should be preferred over ASCII in most cases.

Depending on the specific requirements, several options for file format exist:

- Custom binary formats maximize transfer speed and minimize file size. The drawback is that each client needs to be able to decode the binary data stream. The binary data is not standardized and is not human readable.

- Tables of comma separated values (CSV) contain columns of human readable values. Applications like Matlab and many computer languages have good interfaces to read CSV data. CSV is best suited for the efficient processing of simple tabular data. Problems can arise if the data itself contains control characters (e.g., commas, tabs or newline characters).

- XML is well suited for more complex structured data. It is based on international standards and can contain text in any human language. XML is based on Unicode. Compared to CSV and binary formats it contains more redundancy. Its complex parsing mechanisms offer additional possibilities but cost computing performance. XML is a good choice if size and speed are not critical factors and standardization, openness and comfortable handling of more complex data structureare potentially important

Internet Technologies

With the Internet's triumph in everyday life, there is so much momentum behind the basic Internet technologies and standards that they can hardly be overlooked in process automation. People are used to the Internet way of doing things and powerful tools and techniques exist that can be reused for process automation.

HTML

The HyperText Markup Language (HTML) was created by the founders of W3C and is now a standard, ISO/IEC 15445. HTML allows users to specify web-based user interfaces. One example where HTML can be used in process automation is in configuring GUIs (graphical user interfaces) for smart devices, which might not have their own display. Many of today's sensors contain a microcomputer to add features like calibration and diagnostics, but adding a dedicated display device would be too expensive. In such a case, the device can implement a small web server and offer the user interface via HTML over the network. Another example is thin-client user interfaces making process information available on every standard computer connected to the network.

HTTP

The Hypertext Transfer Protocol (HTTP) is a communications protocol for the transfer of information over computer networks. It has been jointly developed

2.3 Industry Standards and Open Interfaces

by IETF (Internet Engineering Task Force) and W3C. It is a generic, stateless protocol that can be used for many tasks beyond its use for hypertext, such as name servers and distributed object management systems, through the extension of its request methods, error codes and headers (see http://tools.ietf.org/html/rfc2616).

Web Services

The programmatic interfaces that enable application-to-application communication with standard web techniques are referred to as Web Services (see www.w3.org/2002/ws). Techniques like RMI, CORBA, and Distributed Component Object Model (DCOM) had similar goals, but never gained broad acceptance. In general, information technology web services are already widely used by companies like SAP (Netweaver™), IBM, Microsoft, Google, Amazon, and many others. In process automation, web service plays an increasingly important role in standards like OPC UA and ISA-95. Many products and project-specific solutions in process automation use web services. Given the general trend of rapid adoption of web service technology, it is very likely that they will play an important role in future process automation.

Many web services are based on the SOAP standard (see www.w3c.org/TR/soap). SOAP (originally, Simple Object Access Protocol) is a lightweight protocol intended for exchanging structured information in a decentralized, distributed environment. An alternate method of implementing web services was proposed by Fielding (Fielding 2000). These so-called REST (Representational State Transfer) services are easier to implement than those based on SOAP and are preferred by important web service providers like Amazon, Google, and Yahoo.

Web services can be used to implement a service-oriented architecture (SOA). An SOA infrastructure allows loosely coupled components to exchange data with one another as they participate in a business process.

SNMP

The Simple Network Management Protocol (SNMP) was defined by the Internet Engineering Task Force and is used in network management systems to monitor network-attached devices for conditions that warrant administrative attention. It consists of a set of standards for network management, including an Application Layer protocol, a database schema, and a set of data objects.

SNMP can be used to supervise Industrial Ethernet components like routers and switches. CPAS workstations running Microsoft Windows can be supervised with SNMP also.

Microsoft COM

Microsoft Component Object Model (COM™) technology was published in 1993 and enables software components to communicate with each other. COM is a core part of current Microsoft operating systems like XP and Vista. COM can be used to create re-usable software components, link components together to build applications, and take advantage of Windows services. COM is the technological basis of the successful OPC DA, AE, and HDA standards (see section on OPC).

A COM interface is a collection of logically related methods. The COM Interface is an abstract contract on the functionality a COM server can deliver. It defines which methods need to be available and what parameters the methods must have, but it doesn't say anything about how the methods should be implemented. A COM server is a binary software component that exposes functionality as methods in COM interfaces. It uses exactly the methods with the parameters defined in the Interface and puts a concrete implementation behind. A COM client is an application that calls methods of a COM server (Figure 2–16). A COM server can run:

- In the process of the COM client
- In a separate executable on the same machine as the COM client
- In an executable on a different machine (DCOM is required)

COM has been superseded by the .NET™ Common Language Runtime, which provides bi-directional, transparent integration with COM.

COM is available on Microsoft operating systems only, but can be added to other operating systems (e.g., Linux) with the help of third-party add-on packages. COM remains a viable technology with an important software base.

2.3 Industry Standards and Open Interfaces

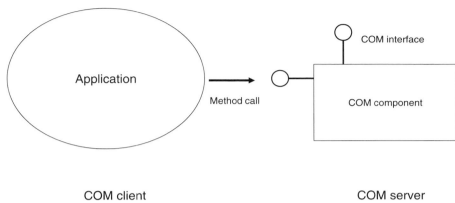

Figure 2–16 Client Application Using COM Component

DCOM

Distributed Component Object Model (DCOM) is an extension of COM for communication among software components distributed across networked computers. The functionality offered by DCOM includes:

- **Authorization**—Authorization defines who is permitted to access which service. The DCOM authorization model is very complex. In practice, many engineers were not able to configure DCOM properly and disabled security more or less to get it working. This, of course, means a security hole.

- **Marshalling**—serializing and deserializing the arguments and return values of method calls.

- **Distributed garbage collection**—ensuring that references held by clients of interfaces are released when, for example, the client process has crashed, or the network connection has been lost.

DCOM was originally made for office applications. Typical requirements from process automation like fast and reliable failover were not sufficiently covered. PAS vendors who decided to build their system on DCOM had to either modify DCOM or to work around the inherent problems. Several products exist on the market that replace the DCOM network communication between a COM client and server with a more robust proprietary communication. DCOM was a constant source of trouble and an important reason why Microsoft developed the .NET technology to succeed COM.

ActiveX

ActiveX controls are COM components with a user interface. ActiveX is an important technology used to plug-in GUI components into a container framework. An ActiveX component needs to implement several interfaces which control the communication between container and component. ActiveX is used in the FDT standard. See section on Field Device Tool (FDT) in this chapter.

ActiveX components run on Microsoft operating systems only. Distributing binary components over the network has proven to be non-secure. Today more secure techniques to distribute GUI components over the web, including Ajax, Flash, and Silverlight™, are available.

SQL

Structured Query Language (SQL) is a database computer language designed for the retrieval and management of data in relational database management systems (RDBMS). ISO/IEC 9075 defines the SQL language. Popular SQL databases are available from Oracle, Microsoft, IBM, and Sun. Many vendors have added proprietary extensions to SQL.

SQL SELECT statements allow to formulate queries and to fetch all matching database rows (Figure 2–17). RDBMS are well suited to storing large amounts of tabular data. RDBMS are used in many CPAS solutions. Sometimes a specialized application programming interface (API) exists, but often data is directly accessed with SQL.

```
SELECT *
FROM alarms
WHERE priority < 2
ORDER BY timestamp
```

Figure 2–17   SQL Query Asking for All Alarms with Priority Below 2

Process Automation Standards

In the following section, standards are described that were specifically developed for process automation needs. Many of the standards are based on the more general standards described in the previous section.

Digital Fieldbuses

Traditionally, field devices have been connected with analog 4–20 mA cables to the PAS. Special I/O cards have then translated the signal from analog to digital. While analog connection is still an option, digital fieldbuses are rapidly gaining popularity. A fieldbus is a digital data bus for communication with industrial control and instrumentation devices, such as transducers, actuators, and controllers. Important fieldbuses in process automation are PROFIBUS, FOUNDATION™ Fieldbus, and HART™.

PROFIBUS

PROFIBUS DP (Decentralized Peripherals) is used to operate sensors and actuators via a centralized controller in production technology. PROFIBUS PA (Process Automation) is used to monitor measuring equipment via a process control system in process engineering. PROFIBUS is standardized in IEC 61158/IEC 61784 (see www.profibus.com).

FOUNDATION Fieldbus

FOUNDATION Fieldbus provides a communications protocol for control and instrumentation systems in which each device has its own "intelligence" and communicates via a digital, serial, two-way communications system. FOUNDATION Fieldbus H1 is designed to operate on existing twisted-pair instrument cabling with power and signal on the same wire. Fiber-optic media is optional. FOUNDATION High-Speed Ethernet (HSE) is ideally suited for use as a control backbone. Running at 100 Mbps, the technology is designed for device, subsystem, and enterprise integration (see www.fieldbus.org).

HART

Highway Addressable Remote Transducer (HART) is a master-slave field communications protocol. The majority of analog field devices installed in plants worldwide are HART-enabled. The HART protocol makes use of the Bell 202 Frequency-Shift Keying (FSK) standard to superimpose digital communication signals at a low level on top of the 4–20 mA analog signal. The HART protocol communicates at 1200 bps without interrupting the 4–20 mA signal and allows a host application (master) to get two or more digital updates per second from a field device (Figure 2–18). As the digital FSK signal is phase continuous, there is no interference with the 4–20 mA signal (see www.hartcomm2.org). The

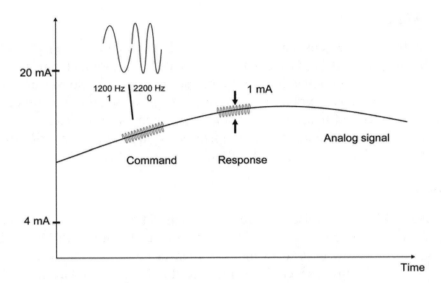

Figure 2–18   HART Protocol

HART protocol was originally developed by Fisher-Rosemount to retrofit 4–20 mA analog communication installations with digital data communication. Today the protocol is owned by the non-profit HART Communication Foundation. To use the protocol, the I/O boards need to be HART-enabled.

WirelessHART is designed to address the needs of the process industry for simple, reliable, and secure wireless communication in industrial plant applications.

Field Device Configuration

Larger plants can have several thousands of digital field devices. The previous generation of analog field devices required physical access to the device (which might be miles away) for calibration and diagnosis. Digital field devices offer much more convenience as many such management tasks can be done remotely; however, management of a large number of different devices from different vendors still remains a challenge. Standards that help to manage different devices with the same tools support efficient engineering and maintenance.

FDT/DTM

Field Device Tool (FDT) technology standardizes the communication interface between field devices and systems. The key feature is its independence from

2.3 Industry Standards and Open Interfaces

the communication protocol and the software environment of either the device or the host system (see www.fdt-jig.org). FDT allows any device to be accessed from any host through any protocol. A Device Type Manager (DTM) is a device driver usually provided by the device manufacturer. DTMs can range from a simple graphical user interface (GUI) for setting device parameters to a highly sophisticated application capable of performing complex real-time calculations for diagnosis and maintenance purposes.

EDDL

Electronic Device Description Language (EDDL) is a language for describing the properties of automation system components. EDDL can describe:

- Device parameters and their dependencies
- Device functions, for example, simulation mode and calibration
- Graphical representations, for example, menus
- Interactions with control devices
- Graphical representations
- Persistent data storage allows devices to store data in a host application without requiring to recognize conventions for saving the data under the host system.

EDDL is applicable to various communication protocols (see www.eddl.org).

FDI

Future Device Integration (FDI) will use a subset of the OPC UA technology within a client-server architecture. FDI will combine the advantages of FDT and EDDL technologies. The planned FDI architecture will be developed with the following guidelines:

- a client-server architecture
- platform and operating system independent
- host system independent
- compatible with existing EDDL- and DTM-based device descriptions

- applicable to any field device communication technologies
- applicable for hierarchical and heterogeneous network topologies
- an open specification and become an international standard

OPC

In 1996, the first OPC standard (Object-Linking and Embedding [OLE] for Process Control) was developed by several automation suppliers together with Microsoft (Iwanitz and Lange, 2006). The standard is based on Microsoft COM and describes how applications should exchange real-time data on a Microsoft platform. As OPC was a good solution to an urgent problem, the standard was rapidly supported by more and more automation system and product vendors. The OPC Foundation (www.opcfoundation.org) has more than 400 members worldwide. The first standard (originally called the "OPC Specification") is now called the "OPC Data Access Specification," or OPC DA. Many people still mean the OPC DA specification when they speak about OPC.

The next-generation OPC standard, called OPC Unified Architecture (UA), will be independent from Microsoft technology. The first parts of the OPC UA standard were published in 2006 (Mahnke et al., 2009). As OPC UA will play a significant role in the future, it is described in Section 2.4. The existing OPC DA, AE, and HDA (described later in this chapter) are now called "OPC Classic." OPC Classic servers work only on a Microsoft Windows operating system.

In the era before OPC, every application that needed to access data from some external device or system had to use specific drivers and protocols (Figure 2–19). This meant a lot of expensive and error-prone custom development work. With OPC, an OPC server can be provided for the device or system. In most cases, the OPC server is provided by the vendor of the device or system, but it can also come from a third-party software vendor. The OPC server can be accessed with standardized methods as specified in the various OPC standards. This means a client application that is able to access OPC can immediately communicate with all devices and systems providing an OPC server. While the original idea behind OPC was to provide standardized communication between hardware devices and applications, it is now also being used as an interface between applications.

The term "OPC server" doesn't mean that it has to run on a dedicated server machine. Several OPC servers might run on the same machine, and OPC clients and servers can well be running on the same machine.

2.3 Industry Standards and Open Interfaces

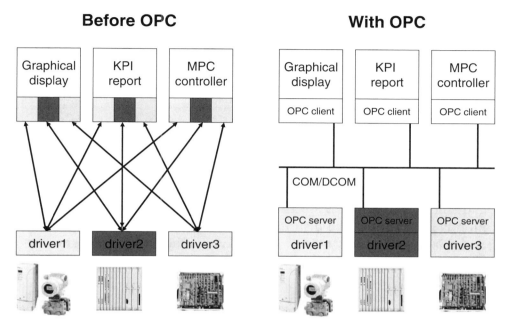

Figure 2–19 Applications Communicating with Devices with and without OPC

OPC DA

OPC Data Access provides the baseline functionality for reading and writing data from various networked devices via a standard set of interfaces. An OPC DA server contains a flat or structured list of OPC items. Usually these items correspond to tags or I/O points. OPC clients can browse the OPC server and see what items are available. The primary intent of OPC Data Access is to provide the interfaces for data acquisition in support of a vertical architecture (serve data from a device to a client application on a higher level computer) (OPC Foundation Data Access Custom Interface Standard V3.00). An OPC items contains a timestamp and a flag for data quality. Various ways of subscribing to OPC items are available, for example, asynchronous or cyclical.

OPC AE

The OPC Alarms & Events specification accommodates a variety of applications that need to share alarm and event information. In particular, there are multiple levels of capability for handling alarm and event functionality, from the simple to the sophisticated (OPC Foundation Alarms and Events Custom Interface Standard V1.1).

Alarm and event engines today produce an added stream of information that must be distributed to users and software clients that are interested in this information. In keeping with the desire to integrate data at all levels of a business, alarm and event information can be considered to be another type of data. This information is a valuable component of the information architecture outlined in the OPC Data Access specification.

OPC HDA

OPC Historical Data Access (OPC Historical Data Access Specification Version 1.2) expands the capabilities of OPC DA, which is focused on current data, to timelines of historical data. Historical engines produce an added source of information that must be distributed to users and software clients that are interested in this information. OPC HDA provides the capability to augment or use existing historical solutions with other capabilities in a plug-and-play environment.

IEC 61131-3

In the past, each PAS supplier created a different programming language for their system. This meant that programs were locked to the systems of that vendor and software developers had to be retrained for each different system. The international standard IEC 61131-3 defines programming languages for programmable controllers. Although IEC 61131-3 programs usually cannot be directly transferred between systems from different suppliers, it greatly supports the transfer of training and know-how between the systems. IEC 61131-3 is described in more detail in Section 3.2.

IEC 61850 Communication Networks and Systems for Power Utility Automation

IEC 61850 is based on Ethernet technology and defines the communication between intelligent electronic devices. It simplifies the integration of electrical components from different manufacturers in power plant automation systems and at the same time reduces the costs of operation and maintenance in these plants. IEC 61850 does not replace process automation fieldbuses; however, it can eliminate the many different serial interfaces in the area of the electrical systems.

2.3 Industry Standards and Open Interfaces

In the past, power plant control systems were connected to the control system via conventional cabling or serial interfaces and gateways. Because of the limitations of these interfaces, a separate control system was often provided for the power plant's electrical systems. Today's state of the art allows the automation of both the process and electrical systems to be combined into one uniform system in power plants. Plant operators and electricians can work with the same control system and use the same graphical displays. The uniformity of the presentation of information and operating procedures improves the quality of the operational processes. With the appropriate authorization, the electrical systems can be operated from any power plant control system workplace.

IEC/EN 61508 Safety Standard for Safety Instrumented Systems

This standard formalizes a systematic approach to the Life Cycle Safety of Safety Instrumented Systems (SIS). Systems like these need to be maintained to be sure of a certain safety level during operation. It is concerned specifically with Electrical/Electronic/Programmable Electronic Safety-Related Systems (E/E/PESs). IEC/EN 61508 provides guidelines to classify these systems by Safety Integrity Level (SIL). Four SILs can be defined according to the risks associated with the system's requirements, with SIL4 (highest reliability) being assigned to the highest risks. The standard adopts a risk-based approach to calculate the required SIL, which represents the Probability of Failure on Demand of the target system (see Section 2.8).

IEC 62264 (ANSI/ISA-95) Enterprise-control System Integration

IEC 62264 is based on ANSI/ISA-95. It defines the interfaces between enterprise activities and control activities. The standard provides a standard terminology and a consistent set of concepts and models for integrating control systems with enterprise systems that will improve communications.

- Part 1: Models and terminology
- Part 2: Object model attributes
- Part 3: Activity models of manufacturing operations management

The interface between manufacturing control functions and other enterprise functions is based upon the Purdue Reference Model for CIM (see Section 5.6 for a detailed discussion).

B2MML

As was mentioned previously, the Business To Manufacturing Markup Language (B2MML) is an XML implementation of the models defined in the ISA-95 series and provided by the Organization for Production Technology (www.wbf.org). B2MML consists of a set of XML schemas written using the XML Schema language (XSD) that implement the data models. B2MML is a complete implementation of ISA-95 as documented in the completeness, compliance, and conformance statement, which is part of the B2MML download. Any company may use B2MML royalty-free provided credit is given to the Organization for Production Technology.

IEC 61512 (ISA-88) Batch Control

The ISA-88 (sometimes called S88) series of standards was first published in 1995 and was later adopted as IEC 61512. The current parts include:

- Part 1: Models and terminology
- Part 2: Data structures and guidelines for languages
- Part 3: General and site recipe models and representation
- Part 4: Batch production records

ISA-88 provides a standard terminology and a consistent set of concepts and models for batch manufacturing plants and batch control. It can be applied regardless of the degree of automation and provides a consistent vocabulary, design process, and representations for Process Control and Logical Control. While ISA-88 was originally intended for batch processes only, it can also be used for managing production targets, process control, and procedural control and reporting of continuous processes (see Section 4.2 for more details).

References

Fielding, R.T. "Architectural Styles and the Design of Network-based Software Architectures." Irvine: University of California, 2000.

Iwanitz, F. and Lange, J. "OPC: Fundamentals, Implementation, and Application." Heidelberg: Hüthig, 2005.

Mahnke, W., Leitner, S.H., and Damm, M. "OPC Unified Architecture." New York: Springer, 2009.

OPC Foundation, *OPC Foundation Data Access Custom Interface Standard V3.00*

——— *OPC Foundation Alarms and Events Custom Interface Standard V1.1.*

——— *OPC Historical Data Access Specification Version 1.2*

Skonnard, A., Gudgin, M. "Essential XML Quick Reference." Upper Saddle River: Addison-Wesley, 2001.

Unicode Consortium, The "The Unicode Standard, Version 5.0." Upper Saddle River: Addison-Wesley, 2006.

2.4 OPC Unified Architecture

Wolfgang Mahnke

Motivation

Classic OPC, with its flavors Data Access (OPC DA), Alarms and Events (OPC A&E) and Historical Data Access (OPC HDA), is well accepted and applied in industrial automation (see Section 2.3). There are over 22,000 OPC products on the market from more than 3500 companies, and all major automation vendors provide OPC solutions (Burke 2008). Thus, almost every system targeting industrial automation implements Classic OPC.

OPC Unified Architecture (OPC UA) (Mahnke et al., 2009) is the new generation of specifications provided by the OPC Foundation and was developed in 5 years by more than 30 vendors. The main goal of OPC UA is to keep all the functionality of Classic OPC and switch from the retiring Microsoft COM/DCOM technology used in Classic OPC to state-of-the-art web services technology. By using web services technology, OPC UA becomes platform-independent and thus can be applied in scenarios where Classic OPC cannot be used today. OPC UA can be seamlessly integrated into manufacturing execution systems (MES) and enterprise resource planning (ERP) systems running on Unix/Linux systems using Java, as well as running on controllers and intelligent devices having specific real-time-capable operation systems. Of course, OPC UA still fits very well in the Windows-based environments where Classic OPC lives today—suiting Microsoft's Windows Communication Foundation, which is based on web services as well.

OPC UA has to fulfill and improve on the non-functional requirements of Classic OPC, providing robust, reliable and high-performing communication suitable to the automation domain. Learning from OPC XML-DA, the first approach of the OPC Foundation to provide XML-based web services, OPC UA supports binary encoding for high-performing data exchange. It has built-in handling of lost messages etc., providing reliable communication. In addition, security is built into OPC UA, an advantage as security requirements

2.4 OPC Unified Architecture

become more and more important in environments requiring the accessing of factory data from office networks.

OPC UA unifies the different specifications of Classic OPC, providing a single point of entry into a system for current data and alarms & events, together with the history of both. There is a single, small set of generic services providing access to all those information.

Whereas Classic OPC has a very simple meta model providing tags in a simple hierarchy, OPC UA provides a rich information model using object-oriented techniques. In OPC UA it is not only possible to offer a measured value and its engineering unit, but also to indicate that it was measured by a specific type of temperature sensor.

This information would be helpful in typical scenarios of Classic OPC. For example, a graphical element of an operator workstation can be defined against the temperature sensor type and used in all occurrences of such a device in the system. In addition, this information can also be used in a broader set of applications in the area of MES and ERP systems, helping to integrate the data without the need to exchange tag lists that provide the semantic content of the tags. OPC UA is designed to allow a rich information model, but it is not required to provide that information. An OPC UA server can still serve information in a manner similar to an OPC DA server today, but it can also provide much more information.

Information modeling is a big advantage of OPC UA compared to Classic OPC, and provides a lot of opportunities. OPC UA defines a simple set of base types that can be extended by information models, either application- and vendor-specific models or standardized models. The idea behind this is OPC UA specifies HOW data is exchanged and standard information models specify WHAT information is exchanged.

There are already several activities ongoing to define standard information models based on OPC UA. There is a draft version of a standard information model describing devices (UA Devices) and based on that, a standard information model for analyzer devices (UA Analyzer) specifying several classes of analyzer devices and their common parameters and characteristics. Using standard information models lifts interoperability to the next level. It not only allows interoperable data exchange but also uses interoperable models. This can, in the long term, drastically reduce engineering costs when integrating systems using products from different vendors.

OPC UA scales very well in several directions. It allows OPC UA applications to run on embedded devices with very limited hardware resources, as well as on very powerful machines like mainframes. Of course, servers running in such different environments will typically not provide the same information. The server on the embedded device will probably not provide a long history of the data and will only support a few clients, whereas other servers may provide the history of several years while supporting thousands of clients.

Also, the information modeling aspects of OPC UA are scalable. A server can provide a very simple model, similar to Classic OPC, up to highly sophisticated models providing high-level meta data on the provided data. A client can just ignore this additional information and provide a simple view on the data, or make use of the meta data provided by the server.

Overview

To fulfill the requirements mentioned in the previous section, OPC UA defines two main pillars as the foundation of interoperability: the communication infrastructure and the OPC UA meta model (see Figure 2–20). The communication infrastructure defines how information is exchanged, and the meta model is the foundation for specifying what information is exchanged.

Independent of the communication infrastructure, OPC UA defines a set of abstract services (UA Part 4) that can run on different communication infrastructures and use the meta model (UA Part 3) as a basis for defining appropriate parameters for the services. The base OPC UA Information Model (UA Part 5) provides base types and entry points into the server's address space. On top of the base information model, vendor-specific or standard information models can be built. OPC UA already defines several standard information models for data access (UA Part 8), alarms and conditions (UA Part 9), programs (UA Part 10), historical data (UA Part 11), and aggregate functions (UA Part 13). OPC UA provides mechanisms to support multiple information models in one server. The information about the information models can be read by the services and thus clients only knowing the services are able to access all the information. Of course, clients knowing specific information models can be optimized by making use of that knowledge.

OPC UA is a multi-part specification split into several documents. Some of the documents have already been referenced in the previous paragraph. The first part (UA Part 1) gives an overview and the second (UA Part 2) defines what security aims are addressed by OPC UA. OPC UA addresses security on a technical level by signing and encrypting messages, but not organizational

2.4 OPC Unified Architecture

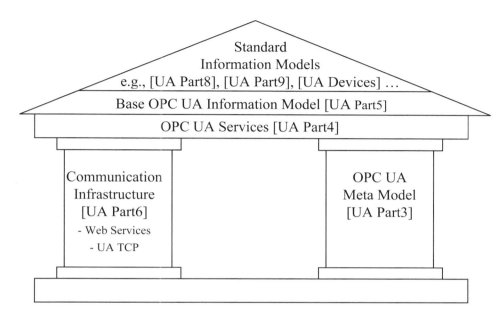

Figure 2–20   OPC UA Overview

issues such as how long a secure password should be and how often it should be changed, etc. This is addressed by other specifications (e.g., the ISA-99 series of standards), and orthogonal to technical aspects like the transport protocol. Parts 1 and 2 are informal documents; the way security is implemented is specified in other parts, including UA Parts 4 and 6.

UA Part 12 defines how "discovery" works (i.e., how clients can find OPC UA servers). The discovery mechanisms also provide information about the capabilities of a server. An embedded server, for example, may not provide historical data. OPC UA defines profiles (UA Part 7) dealing with the fact that different applications have different capabilities. For servers there are currently three base profiles for a standard server, an embedded server, and a low-end embedded server. Each of these profiles defines a minimum set of functionalities a server implementing the profile must support. In addition, servers can implement additional functionality, captured in profile facets. For clients, only facets exist.

The OPC Foundation is currently implementing a compliance program for OPC UA (as well as for Classic OPC) with compliance test tools, interoperability workshops where vendors plug-in their different products, and independent test labs. OPC vendors can get a compliance certificate for their tested products.

In the following section, we will give an overview of how information modeling works in OPC UA and how it is used by standard information models. Afterwards we will investigate details on how the communication infrastructure and services work and will end this chapter by giving an overview of the current status of OPC UA.

Information Modeling

Information Modeling in OPC UA aims at supporting simple information models like the ones used in Classic OPC, as well as more complex information models, providing a rich set of semantic information. This allows for tailoring an application program to an information model and applying it in several places. An example is a graphical element representing a device, displaying its measured values, and allowing the device to be configured. Such a graphic can be applied several times in the same site, or in a library used in several installations.

Providing meta data simplifies the integration of data into other systems such as MES or ERP systems. Here, the integration can be (semi-)automated. In the long term, it can even be used not just to configure existing structures (e.g., configure a device) but also to engineer and configure a system (e.g., by adding devices).

To support all these different use cases, OPC UA uses the following main principles (Mahnke et al., 2009):

- Using object-oriented techniques including type hierarchies and inheritance.

  Typed instances allow clients to handle all instances of the same type in the same way. Type hierarchies allow clients to work with base types and to ignore more specialized information.

- Type information is exposed and can be accessed the same way as instances.

  The type information is provided by the OPC UA server and can be accessed with the same mechanisms used to access instances.

- Full meshed network of nodes allows information to be connected in various ways.

2.4 OPC Unified Architecture

OPC UA allows supporting several hierarchies exposing different semantics and references between nodes of those hierarchies. Thus the same information can be exposed in different ways, providing different paths and ways to organize the information in the same server, depending on the use case.[1]

- Extensibility regarding the type hierarchies as well as the types of references between nodes.

  OPC UA is extensible in several ways regarding the modeling of information (e.g., by subtyping and by adding new types of references).

- No limitations on how to model information in order to allow an appropriate model for the provided data.

  OPC UA servers targeting a system that already contains a rich information model can expose that model "natively" in OPC UA instead of mapping the model to a different model.

- OPC UA information modeling is always done on server-side.

  OPC UA information models always exist on OPC UA servers, not on client-side. They can be accessed and modified from OPC UA clients but an OPC UA client does not have to provide such information to an OPC UA server.

The main concepts of OPC UA regarding information modeling are *nodes* and *references*. There are different classes of nodes, for example, representing objects or variables. A reference connects two nodes with each other, for example, indicating that an object has a variable. As OPC UA allows providing different hierarchies and a full meshed network of nodes, references in OPC UA are typed. The type of the reference defines the semantics of the reference.

Each node has *attributes* describing the node. Some attributes are common to all nodes; other attributes only exist for specific node classes. For example, each node has a NodeId uniquely identifying the node, whereas only type nodes have the attribute IsAbstract indicating whether the type is abstract or not.

The most important node classes are *object, variable,* and *method*. These node classes are used to provide the information on the instances.

---

1. Nodes in an OPC UA address space.

An *object* is used to represent a real object like a device, or an organizational concept like an area. In that way it is used to structure the address space and does not contain data other than those describing the object. An object can contain objects, variables, and methods by referencing them.

A *variable* represents a value like a measured value, a set point or some configuration data. A variable has a specific attribute containing the value called Value. The value can have a specific data type for each variable, specified in other attributes of the variable. Clients typically subscribe to changes of the value attribute of variables or change it by writing it.

A *method* represents something that can be called by the client and is immediately executed by the server. A method has input and output arguments. A method node only represents the interface of a method, not the implementation. Clients can read the meta information of the method and then call the method on the server.

In Figure 2–21, the attributes of objects and variables are shown. There is an object called FT1001 and a variable called DataItem. Details on the attributes can be found in Mahnke et al. (2009). OPC UA defines a fixed set of attributes. If a node needs more information added to it, variables have to be used. Special types of variables are used for this purpose, called *properties*. The variable DataItem has several properties called TimeZone, Definition, and so on.

An *object type* defines the type of an object. All objects are typed and reference an object type node. There is an object type hierarchy having the BaseObjectType as the root. If there is no real type information available, the BaseObjectType should be used.

Variable types define types of variables. OPC UA distinguishes two types of variables: data variables and properties. Properties can be used on each node, containing additional information about the node, whereas data variables represent variables of an object like measured values. Properties are not typed (all referencing the PropertyType) but are defined by the BrowseName, whereas data variables are typed in a data variable type hierarchy.

Data types are represented as nodes in the address space. There is a set of built-in types like string, several integer types, double, etc. In addition, enumeration types and structured types can be defined. Structured types represent complex, user-defined data types in OPC UA. Clients can read the information about the types, including how to handle (encode and decode) them; thus, they are able to deal with those data types. Variables contain the information about the used data type.

2.4 OPC Unified Architecture

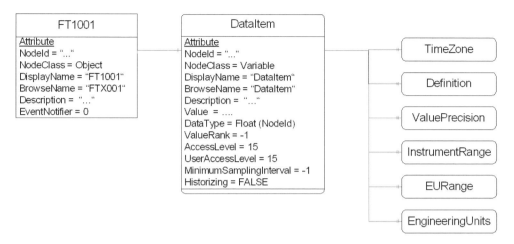

Figure 2–21   Example of an Object and a Variable Including Their Attributes in OPC UA

References are typed. The *reference types* are represented as nodes in the address space, spanning a reference type hierarchy. Clients can use this hierarchy to filter for specific references when browsing and querying the address space.

A *view* is used in potentially large address spaces with hundreds of thousands of nodes to restrict the number of visible nodes and references to a specific task or use case. For example, a maintenance engineer is only interested in information regarding maintenance. A *view* is represented by a node serving as entry point into the view. Afterwards, clients use the NodeId of the view node as a filter parameter when browsing or querying and thus only receive results contained in the view.

Alarms & Events and historical data are seamlessly integrated into OPC UA. The history of current data can be gained from a variable. The AccessLevel attribute not only provides the read and write capabilities of the variable regarding current values but also the history. The Historizing attribute indicates if history is being collected at the moment. Specific objects represent the configuration of how history is collected for a variable.

Alarms & Events can be received by objects having the EventNotifier attribute set to generate events. The attribute also captures whether the history of events is available and can be modified. Configuring event history is represented similar to configuring current data. Specific types of references allow providing the event hierarchy, and specific object types define the types of events and their fields.

Standard Information Models

OPC UA allows providing rich information. However, the concepts of OPC UA allow providing the same information in different ways. In order to increase interoperability on the modeling level and thus simplify the usage, it is possible to standardize how to provide information. Standard information models define standard object and variable types, standard data types and standard reference types. They can also define standard objects and variables as entry points into the address space.

There are activities now taking place to define standard information models based on OPC UA. The OPC Foundation cooperates with MIMOSA (Machinery Information Management Open Systems Alliance), OMAC (The Open Modular Architecture Controls Users' Group), International Society of Automation (ISA), and PLCopen to define standard information models, for example, in cooperation with PLCopen for IEC 61131 languages. Some standard information models are built into OPC UA, such as specific variable types for data access (UA Part 8). In the next section, we sketch how standard information models work, looking at device integration. The ECT (EDD Cooperation Team) has defined an OPC UA information model for devices by defining a base device type and its components (how parameters are organized, etc.) (UA Devices) and a mapping from EDDL to OPC UA using this model. The FDT (Field Device Tool) group has defined a mapping from FDT to OPC UA using this model.[2] Another group used the device model and refined it, defining a model for analyzer devices (UA Analyzer). Here, concrete classes of analyzer devices are modeled with standard parameters.

As shown in Figure 2–22, everything is based on the base information model defined by OPC UA. The devices specification also makes use of the data access model, and the analyzer device specification uses the devices specification as base. On top of this, the vendor-specific information model adds vendor-specific information.

Clients can access all information without the need to know the information models. They can read the information about the types from the server if they want to make use of the type information. Clients knowing the information model can make use of it and be tailored to specific information models. For example, a client can be implemented handling all analyzer devices in a specific way.

---

2. Now both groups are working together on FDI—Field Device Integration—taking OPC UA and the device model as base.

2.4 OPC Unified Architecture

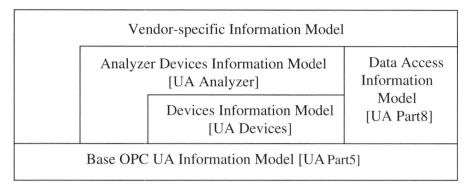

Figure 2–22   OPC UA Standard Information Models

Communication Infrastructure and Services

OPC UA defines the protocol for how data are exchanged. This is in contrast to Classic OPC, where only interfaces have been defined based on Microsoft's COM/DCOM technology. The main reason for this is that Classic OPC had its problems with a technology (DCOM) built for business applications, where it is acceptable that a client may lose its connection to a server for several minutes without noticing. This behavior is not acceptable in industrial automation, and several vendors used their own protocols for remote connections under the COM layer and did not use DCOM. By defining the protocol, the OPC UA specification can control how the data is exchanged and can consider requirements from industrial applications. However, OPC UA is not reinventing the wheel but is using standard technologies like SOAP and WS Secure Conversation wherever this is suitable for the specific requirements of industrial applications.

In order to be as technology independent as possible, OPC UA specifies abstract services in UA Part 4 and the mapping to technologies in UA Part 6. This allows adapting OPC UA to a new technology by adding a new mapping without redefining the services. The OPC Foundation provides stacks for different programming languages implementing the technology binding (see Figure 2–23). All stacks provide a programming-language-specific and non-standardized API that applications are using. The transport implemented by the stack is specified by OPC UA and thus all stacks can communicate with each other. This architecture allows adding technology bindings by just adding the implementation to the stacks without changing the application. It is not needed

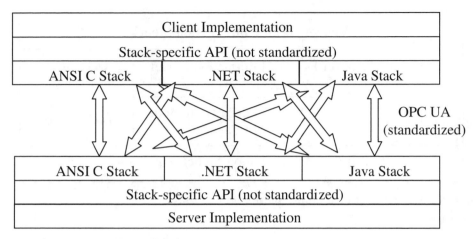

Figure 2–23 OPC UA Architecture

(although it is recommended) to use a stack from the OPC Foundation. For specific technology bindings, it is suitable to generate your own stack (e.g., based on a WSDL [Web Service Description Language] defining the web services).

The technology mappings currently defined are summarized in Figure 2–24. OPC UA supports pure web services with XML encoded data using WS Secure Conversation (left side). This setup may be reasonable in scenarios where the client is a pure web-service-based client generating its stack from a WSDL and not needing fast access to the data (e.g., in the ERP environment).

The other extreme is using the UA TCP protocol based on TCP/IP and binary encoding the data (right side). This setup provides the highest performance by minimizing the protocol overhead and binary encoding the data. Since WS Secure Conversation is defined for SOAP messages, it is adapted to UA Secure Conversation supporting the UA TCP protocol. The middle of the figure shows combinations of both approaches, the left one using binary encoded data for higher performance in the web service world and the right one supporting firewall-friendly access of the high-performing solution by wrapping the messages into SOAP messages.

The abstract services of OPC UA are summarized in Figure 2–25. To establish a connection, a secure channel first has to be created. The secure channel may or may not support security features like signing and encrypting. This depends on the configuration. OPC UA requires compliant applications to implement security but does not force the use of security. There are scenarios where security is not important but performance is critical, and security costs performance.

2.4 OPC Unified Architecture

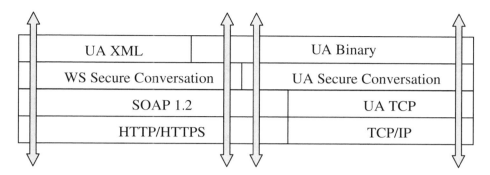

Figure 2–24 OPC UA Technology Mappings

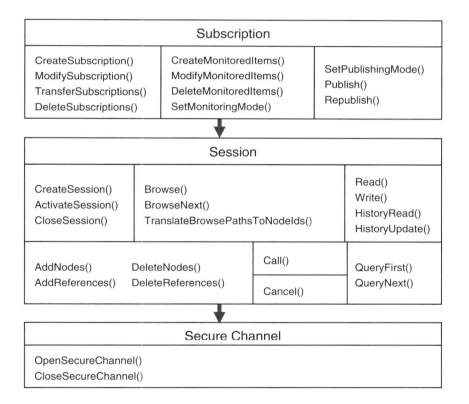

Figure 2–25 OPC UA Services Overview

On a secure channel, a session is established. Using a session, clients can browse and read the address space of the server. They can manipulate the address space by writing values or changing the address space (adding and deleting nodes and references). They can call methods, and in complex address spaces they can query the address space and cancel requests where the result is not needed any more.

On top of a session, subscriptions are used to subscribe to events and data changes. Clients can specify to what data they want to subscribe to by creating monitored items and using the Publish mechanism to receive the data. Details on the services and how they work can be found in Mahnke et al. (2009).

As OPC UA is potentially running on the Internet and using technologies that are not reliable by themselves, the services take care of this by allowing to re-request lost messages, maintaining keep-alive counters, session management, etc.

Status of OPC UA

The first version of the OPC UA specification was released in 2006. This included UA Parts 1–5, followed by some information model extensions (Parts 8, 10 and 11) by January 2007. It did not include the technology mapping (UA Part 6) and thus was not enough to implement products compliant with the specification. However, an early adopter program started even before the first release of the specification, developing the infrastructure supporting OPC UA such as stacks in C, .NET, and Java, Software Development Kits (SDKs) based on .NET, and wrappers and proxies for the Classic OPC specifications. While this work was being finished, an updated release of the published versions and the technology mapping and the profiles was released in the beginning of 2009. The OPC UA specification will be published as IEC 62541. Now everything is in place to implement OPC UA products. The first products were already released before the specification was released.

The compliance program of the OPC Foundation for OPC UA is nearly implemented, guaranteeing interoperable products by compliance test tools, interoperability workshops, and independent test labs.

To summarize, OPC UA is ready to replace Classic OPC, using state-of-the art technology in a reliable, secure and high-performing way and to bring interoperability in automation to the next level by allowing standard information models based on OPC UA.

References

ANSI/ISA–99.00.01–2007 *Security for Industrial Automation and Control Systems: Concepts, Terminology and Models*, Research Triangle Park: ISA, 2007.

Burke, T. "OPC and Intro OPC UA" Presentation at OPC UA DevCon, Munich, 2008.

Mahnke, W., Leitner, S.H., and Damm, M. "OPC Unified Architecture" ISBN: 978-3-540-68898-3, New York: Springer, 2009.

OPC Foundation *Analyzer Devices,* Draft Version 0.30.00, Dec. 2008.

———————— *UA Spec. Part 1 – Concepts,* Version 1.01, Feb. 2009.

———————— *UA Spec. Part 2 – Security Model,* Version 1.01, Feb. 2009.

———————— *UA Spec. Part 3 – Address Space Model,* Version 1.01, Feb. 2009.

———————— *UA Spec. Part 4 – Services,* Version 1.01, Feb. 2009.

———————— *UA Spec. Part 5 – Information Model,* Version 1.01, Feb. 2009.

———————— *UA Spec. Part 6 – Mappings,* Version 1.0, Feb. 2009.

———————— *UA Spec. Part 7 – Profiles,* Version 1.0, Feb. 2009.

———————— *UA Spec. Part 8 – Data Access,* Version 1.01, Feb. 2009.

———————— *UA Spec. Part 9 – Alarms and Conditions,* Draft Version 0.93q, Nov. 2007.

———————— *UA Spec. Part 10 – Programs,* Version 1.00, Jan. 2007.

———————— *UA Spec. Part 11 – Historical Access,* Version 1.00, Jan. 2007.

———————— *UA Spec. Part 12 – Discovery,* Draft Version 1.00.03, Nov. 2007.

———————— *UA Spec. Part 13 – Aggregates,* RC Version 1.0, July 2008.

————————*UA Devices*, Draft Version 0.75, Dec. 2008.

2.5 Dependable Automation

Hubert Kirrmann

Introduction

Dependability is the quality of a system to achieve a predefined behavior in the face of failures of its parts. It encompasses safety, availability, reliability, and maintainability. Dependability is sometimes called "RAMS", for these four factors. Dependability is a major requirement in automation systems. The most important requirement is safety, i.e., ensure that a failure of the control system does not cause harm to people or to the plant. The second most important objective is high availability, i.e., a failure of the control system should interrupt production as little as possible (or even not at all). The third requirement is reliability, which expresses how often the system fails; if redundancy is available, failure of a backed-up part will not immediately translate into a loss of availability, but frequent calls to maintenance are undesirable. The fourth requirement is maintainability, which aims at reducing down time or even keep at zero.

Solutions to guarantee safety and availability are an important part of CPAS offerings. Although operating principles vary, the dependable architectures can be divided into a relatively small number of types. However, it has been recognized that general-purpose solutions are complex and costly, and that fault-tolerant automation systems should be tailored to meet the needs of specific plants. This means that many fault-tolerance mechanisms are not transparent to the application programmer and thus have additional engineering costs. A precise knowledge of the plant is necessary, as one of the most limiting factors is the unreliability of the redundant hardware that was specifically introduced to increase fault tolerance.

Plant Type and Computer Architecture

Plant Types and States

A CPAS contains controllers interconnected with sensors and actors (actuators, pumps, etc.) over communication systems. A component of the automation system can fail in a number of ways, which can be condensed to two classes:

- Integrity breach: the system delivers wrong outputs, either in the value or in the time domain.

- Persistency breach: the system fails to deliver usable outputs.

These properties apply to programmable controllers, communication systems, storage systems and input/output devices. We will only consider controller failures, since they represent the three matters of interest (processing, communication, and storage) for all other subsystems, so the principles apply to the other components in a simplified form.

What happens to the plant in the event of failure depends on the kind of failure and on the plant's characteristics.

The most important issue after a failure has occurred is whether the plant is still safe (i.e., does not enter a state where major damage or injuries occur). The distinction between safe failures and unsafe failures can only be made upon analysis of the plant.

In this chapter, we will not consider intrinsic safety, i.e., damage to the plant resulting from the malfunction of a device, such as burning or generating sparks in an explosive environment.

If the automation system delivers wrong data (integrity breach), some plants will respond to such a breach by some scrap production, causing minor damage, but other plants do not tolerate integrity breaches at all. For instance, an electrical substation will not tolerate that a switch is operated upon a controller failure.

Plants generally incorporate protections (i.e., some redundancy) to mitigate a controller's failure. For instance, upon detection of a controller error, an emergency mechanism can shut down a plant to a safe state, preferably passively. Also, high-level plausibility checks can ensure safety by preventing potentially dangerous operations, without identifying their cause. In the example of the substation, interlocking mechanisms will prevent a switch from being operated upon a controller failure if this would endanger the substation.

If the automation system does not generate usable data (persistency breach), some plants will just suffer from production losses, but others will become unsafe if not controlled. For instance, an airplane can tolerate erratic commands during a few seconds, provided that usable data flow is resumed quickly. For this kind of "plant," which has no safe position, safety is a matter of persistency.

When a protection function fails, a persistency breach implies a safety hazard since the plant is no longer protected and should be shut down if not repaired within a given time, causing production loss and lowering availability.

The important design factor is how long a plant can tolerate a malfunction of its control computer without tripping (going to the safe shutdown state) or without suffering damage. This parameter, called the grace time, is specific to a plant. Typical grace time values are shown in Table 2–1.

When the automation system is fault-tolerant, it can overcome a number of failures before causing a shutdown. Without repair, however, the plant would ultimately go down. In a plant that is operated 24 hours a day, such as a power plant, the reintegration of repaired components without stopping is necessary to ensure availability.

A taxonomy of failure modes related to safety and guidelines is given in different standards, the most common being IEC 61508 "Functional safety of electrical/electronic/programmable electronic safety-related systems" and IEC 61511 "Safety instrumented systems for the process industry sector."

These standards make little reference to implementation, except IEC 61508, which defines the MooN (M out of N) definition that is used in this chapter to describe the structures.

Figure 2–26 shows the states of a typical plant that is controlled by a (redundant) automation system.

Table 2–1 Grace Times

Processing industry:	10s
Chemical processes:	1s
Rail vehicles:	100 ms
Printing machines:	20 ms
Drive-by-wire cars:	5 ms
Electrical substations:	4 ms

2.5 *Dependable Automation*   103

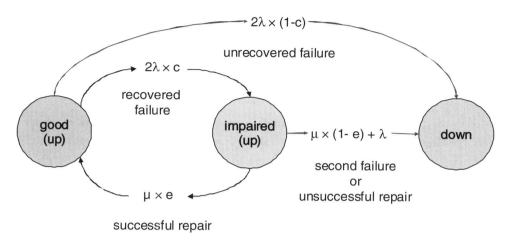

Figure 2–26   Model of a Highly Available Automation System

In this model, we consider a system that includes functional redundancy. Originally, in state "good," all elements are working.

The system leaves this state with a probability 2λ when a part fails, since redundancy normally fails with the same probability as the part it should back-up.

At the first failure, either the failure is handled correctly, or failover (automatic change-over to redundant component) fails, with a probability "c" (coverage) for success. The latter, for instance, could be the case if the failure is a design failure, or a failure that could not be detected in time.

Either a redundancy can take over in time, as expressed by transition $(\lambda - c)$ and the plant continues to operate; or it cannot take over in time to keep the plant producing, as expressed by transition $\lambda - (1 - c)$.

The transition μ indicates a repair. Here also, a repair can be successful with a probability e (the plant operates again at full redundancy) or the repair causes a failure of the plant (assuming on-line repair).

Repair can take place when the plant is shut down on a scheduled basis, causing little production loss. But with continuous plants, on-line repair is a necessary and critical operation.

High Integrity Systems

The first requirement of a controller is to maintain data integrity in the case of failure of one of its parts. The ideal behavior is "fail-silent", in which the controller does not output any data and behaves passively on the communication network. This behavior is sometimes improperly called "fail-safe."

To strive for this behavior, controllers are designed with a number of error detection and fault confinement mechanisms, such as watchdog timers, memory partition, time partition, plausibility checks, memory checks, and arithmetic checks.

Even extensive diagnostics cannot prevent false data from escaping from the controller during a certain time, since diagnostics are not constantly active. Systems incorporating extensive diagnostics are called 1oo1D systems (see Figure 2–27).

To provide complete integrity, it is necessary to use duplication and comparison, a configuration in which each operation of the worker controller is constantly checked by another controller, the checker (see Figure 2–28).

The checker preferably is not identical to the worker, to reduce the probability of systematic errors. This applies especially to software errors.

ABB's AC800M HI, for instance, uses different operating systems on the worker (PM865) and the checker (SM810) to provide high-integrity coverage (see Figure 2–29).

Persistent Systems

Persistent systems rely on a functional redundancy to replace a failed operating part.

Figure 2–30 shows the principal arrangement. A worker controller is backed by a replacement unit with the same functionality; error detectors allow the fail-over logic to choose which unit is to be trusted.

Inserting redundant hardware is not sufficient; this hardware must also be actualized (loaded with the same state) as the unit it replaces, otherwise fail-over will not be bumpless and can result in an integrity breach.

There are two methods for keeping the back-up properly actualized, standby and workby (see Figure 2–31).

In standby, only one unit is working, and it keeps the other actualized by transferring its state over a high-speed link to the back-up, which is passive. The actualization is not continuous, but takes place at determined checkpoints in the execution of the worker. When the error detection signals that the worker has failed, the back-up takes over and restarts itself to the last correctly received checkpoint.

2.5 Dependable Automation

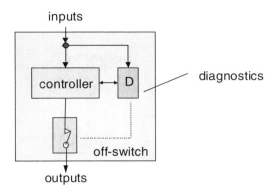

Figure 2–27   Simplex (1oo1D) Controller

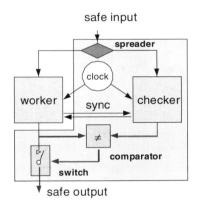

Figure 2–28   Duplication and Comparison (2oo2)

The advantage of standby is that the programming and operation of the worker are not strongly influenced by the presence of the back-up, except for the time it costs to transfer the state changes. The disadvantage is that failover costs some time; typical values are in the range of 10 ms–100 ms.

In workby, both units are synchronized and operate in parallel. This means that, they both receive the same inputs and synchronize their transitions over a high-speed link. Since both units are indistinguishable, the plant can select either one of them as long as the error detectors do not signal an error.

The advantage of workby is that failover is completely smooth. From a reliability point of view, workby allows constantly exercising redundancy and therefore uncovers lurking failures.

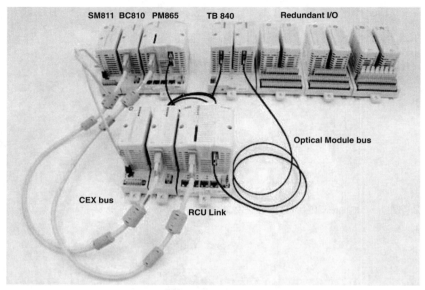

Figure 2–29   High Integrity with the AC800M HI (SIL3 Certified)

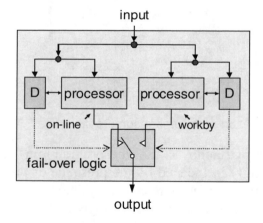

Figure 2–30   Persistent Controllers (1oo2D)

Another advantage is that this configuration can be used for duplication and comparison, and therefore also provides integrity. Signals can be blocked in case of discrepancy as long as one unit is not declared as failed.

The disadvantage of workby is that the controllers must closely synchronize their operations, and in particular their interrupts and clocks. This has little impact on speed, but a greater impact on the operating system.

2.5 Dependable Automation

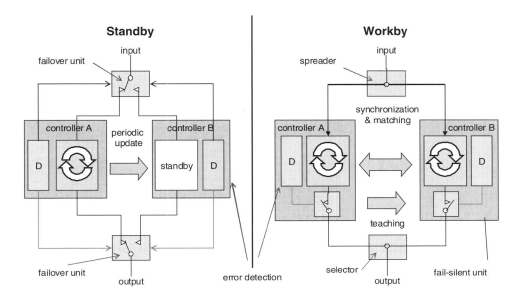

Figure 2–31   Standby and Workby

Most automation systems operate on the standby principle since a simplex system (no redundancy) can evolve into a duplex (redundant) system without major reengineering, and redundant parts of the plant can be connected easily to non-redundant ones.

For instance, the ABB 800xA family extensively uses this method for controllers, communication systems and I/O, as Figure 2–32 shows for an automation system consisting of duplicated and simplex parts. The simplex part can nevertheless provide redundancy through duplicate I/O devices.

Both standby and workby share a common challenge: reintegration of repaired parts in a running system for 24/7 plants. The running machine must "teach up" the replacement until the replacement is able to operate as a back-up again. This creates a peak load on the controller that either slows it down during a certain time or costs a reserve of computing power, just for the purposes of this infrequent situation.

High Integrity and Persistent Systems

When neither wrong data nor loss of data is tolerable, error masking is needed. Masking is partially possible with a workby configuration, since in the case of a

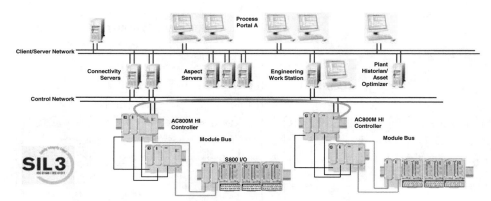

Figure 2–32 800xA System with Redundant and Simplex Parts

discrepancy between worker and co-worker, the output signals can be blocked until one unit is identified as failed. Persistency is, however, lost while the diagnostics are being run.

The classical solution to both integrity and persistency, also called "reliable" controllers, is triplication and voting (Triplicated Modular Redundancy, TMR), a method used since the dawn of computing, for instance in the Saturn V moon rocket.

Here, three units operate in workby and a voter takes the majority (see Figure 2–33).

TMR has the advantage of masking every failure.

TMR's disadvantages are that hardware costs are high (triplicate processors, plus voter, synchronization links and spare resources). The close synchronization of three controllers requires a complicated consensus and presents technical challenges when a unit has to be replaced and reintegrated. And, of course, a TMR controller fails three times (at least) more often than a simplex one, although the consequences of a failure are well controlled.

TMR requires special development. For instance, ABB August is a separate line of products that specifically addresses the high-end fault tolerance market, with different processor types and development environments that consider triplication. An example is ABB's Triguard (see Figure 2–34).

To overcome the need for special development, one can also pair high-integrity controllers into a Quad. The pairs can be operated in workby or in standby (see Figure 2–35).

An example is ABB's Safeguard (see Figure 2–36).

2.5 Dependable Automation

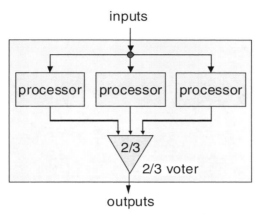

Figure 2-33   TMR or 2oo3v

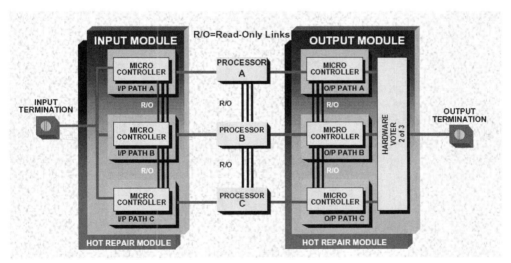

Figure 2-34   ABB's Triguard

The ownership costs of fault-tolerant systems are high and their use is only justified when the consequences of a failure are not bearable. A cost-effective fault-tolerance approach first looks at how tolerant the plant is to failures rather than simply applying massive redundancy.

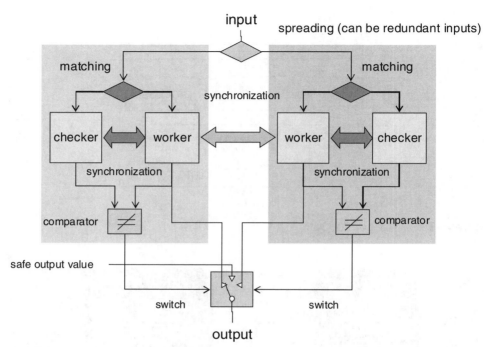

Figure 2–35   Quad 2oo4 System

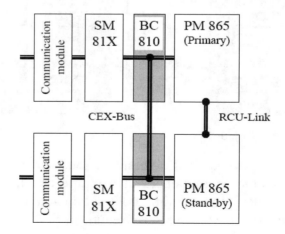

Figure 2–36   ABB's Safeguard System

Bibliography

Grosspietsch, K-E. and Kirrmann, H. "Fehlertolerante Steuerungs- und Regelungssysteme." Automatisierungtechnik at 8 (2002).

Hölscher, H. and Rader J. *Microcomputers in Safety Technique, An Aid to Orientation for Developer and Manufacturer.* TÜV, Rheinland, Germany: 1986.

International Electrotechnical Commission, *IEC 61508: Functional Safety of Electrical/Electronic/Programmable Electronic Safety-related Systems, Parts 1 to 4.* Geneva: 2006.

——— *IEC 61511-1 Functional Safety—Safety Instrumented Systems for the Process Industry Sector—Part 1: Framework, Definitions, System, Hardware and Software Requirements.* Geneva: 2008.

Siewiorek, D.P. and Swarz, R.S. *The Theory and Practice of Reliable System Design.* Digital Press, ISBN: 0-932376-13-4 Bedford, 1982.

2.6 Scalability and Versatility

Martin Hollender

Introduction

Using a scalable and versatile CPAS possessing multi-functional capabilities, adaptability to changing requirements, openness, and maximum availability is paramount to success. It allows the CPAS to grow with the needs of its users, protects investments, and ensures openness for future enhancements.

The scalability of a system has been defined as its ability to either handle growing amounts of work in a graceful manner, or to be readily enlarged (Bondi, 2000). Versatility means that the same system can be used to automate a broad spectrum of processes. These processes have different automation requirements depending on:

- Industry
    - Oil and gas
    - Power
    - Pulp and paper
- Size
    - Number of I/O points
    - Number of control loops
    - Number of users
- Performance
    - Storage resolution and horizon
    - Sampling rate
    - Data throughput

2.6 Scalability and Versatility 113

- Complexity
  - Control algorithms
  - Interfaces between modules
- Available budget
- Many other factors

It is obvious that a small brewery has different automation requirements than a large refinery. In one area, users want a cheap and easy-to-engineer solution, whereas in other areas much more focus is put on safety and availability.

If a CPAS is tailored for a specific use case, it can be highly optimized for that particular case; however, it will probably be restricted to a small niche.

On the other hand, a one-size-fits-all CPAS also means a compromise; therefore, it is very important that the CPAS can be customized for the needs of different industries. The aim is to make the reusable common CPAS core as big as possible, so that relatively little customization work remains.

Good examples are safety instrumented systems (SIS), which need to comply with the IEC 61508 and IEC 61511 standards (see Section 2.8). These include safety critical applications such as emergency shutdown systems, fire and gas systems, and burner management. In the past, SIS were implemented with completely separate systems. Modern CPAS can offer a single unified, high-integrity system architecture covering both SIS solutions and standard process control. This unified, high-integrity architecture reduces duality and the associated life-cycle costs of maintaining separate process control and SIS systems. Project engineering, training, operations, maintenance, and spare parts can be optimized through the use of the common architecture.

The Value of Scalability and Versatility

The main reason to design for scalability and versatility is reduced cost and effort. Major changes in business goals will inevitably occur, but allowing for easier, iterative updates and additions can reduce the cost and effort of those changes.

Standardization

Many companies operate a large portfolio of processes with very different requirements. For example, petrochemical plants may include a power plant to produce electricity. A chemical company might operate both small batch processes and large continuous processes. The option of implementing the same CPAS for different use cases has the following benefits:

- Less training is required because know-how and expertise can be transferred between installations.
- Fewer spare parts need to be managed and CPAS installations can be maintained by the same technicians and experts.
- Custom solutions developed for one specific case can be more easily transferred to other installations if the same CPAS is being used.

Potential for Growth

It is often a good strategy to start a new product small, with limited risk. If the new product is successful, the demand rises and its production needs to be increased. It would mean a lot of rework and extra cost if bottlenecks should force the producer to change to a different CPAS each time production is increased.

Many companies want to start simple and add more advanced automation functionality later. For example, they might want to add an Asset Management System or an Information Management solution. It is not acceptable that adding such new functionality should require a complete rework of the automation system.

Mass Production Reduces Cost

The main cost blocks of a CPAS vendor include:

- Development
- Testing
- Marketing
- Shipping
- Training
- Maintenance and upgrade

2.6 Scalability and Versatility

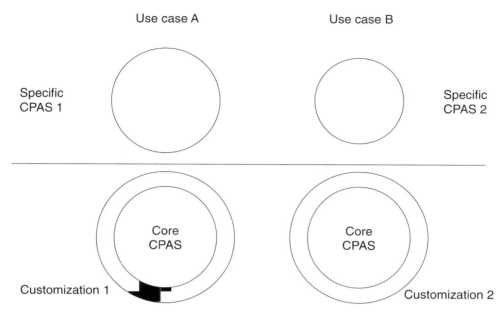

Figure 2–37 Customization vs Specialization

If several completely separate CPAS special-purpose implementations need to be engineered, this means lots of duplicate work and therefore much higher cost. As a consequence, the vendor cannot spend as many resources for each specific CPAS.

To respond to different requirements from different industries, customization is inevitable (see Figure 2–37). To avoid non-value-adding, duplicate work as much as possible, the common core CPAS should be maximized and the required customizations be minimized.

Design for Scalability

Central bottlenecks like, for example, the maximum number of I/O points, the maximum number of clients, or the throughput, may impose limits to the scalability of a CPAS. Such bottlenecks need to be avoided by using a modular design, where modules can be added on demand to increase performance and/or functionality. Central resources like network connections need to be dimensioned in a way that they can support the biggest systems.

The spectrum between small/simple and large/complex automation projects is rather broad and difficult to satisfy by the same CPAS architecture. To

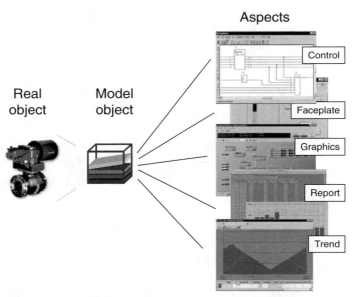

Figure 2–38 The Aspect Object Concept Allows Functionality to be Added on Demand

permit the use of a single basic architecture, in spite of the additional cost, CPAS vendors can offer less expensive, slimmed-down entry-level offerings for small/simple projects. Even if such solutions are more expensive and complex than dedicated entry-level automation systems, their scalability pays off when seen in a broader context.

Design for Versatility

Versatility calls for generic, configurable solutions. Dedicated solutions for specific use cases should be avoided. Instead, the core CPAS needs to provide a generic framework that can be customized for different needs. For example, the workplace layout should not be optimized for one specific industry (e.g., power plants) because all other industries would have difficulties in adapting to such a layout. Instead the core CPAS needs to provide configurable mechanisms so that customized layouts can be provided for different industries.

The Aspect Object Concept (Figure 2–38) is explained in Section 2.2. It allows adding specific aspects on demand; for example, lets assume that a plant wants to add asset monitor functionality to their already existing solution. The Aspect Object Concept allows adding asset monitor aspects to existing objects.

2.6 Scalability and Versatility

In many cases, this can be done in a few central places and is then automatically inherited to all relevant objects. The asset monitor aspects can fetch context information from other aspects on the same object.

References

Bondi, A. "Characteristics of Scalability and Their Impact on Performance." ACM *Proceedings of the 2nd International Workshop on Software and Performance*, pp 195–203. Ottawa, Canada: 2000. ISBN: 1-58113-195-X.

2.7 IT Security for Automation Systems

Markus Brändle and Martin Naedele

Introduction

A key characteristic of a CPAS is its integration into a web of communication connections with other IT systems, both inside and outside the enterprise. These communication connections serve to improve the visibility of the actual production status in enterprise planning systems, to avoid repeated entry of data, such as orders and staff hours in different systems, and also to access systems for diagnosis, configuration, and engineering remotely to speed up reactions and to save travel costs.

Today, these communication connections increasingly use open and standardized protocols and technologies and commercial-off-the-shelf (COTS) applications, as well as "Internet technologies" and the Internet itself. The deployment of these technologies in a process control system and the connection to a potentially untrustworthy outside world introduce IT security as a new concern for process control systems.

Thus, IT security has become one of the dominant topics with respect to public and governmental attention to the process control industry over the last few years. While there is a consensus in the process control industry that a threat of attacks exists and needs to be addressed, there does not yet seem to be a shared understanding of how big the risk really is. Nevertheless, protection against electronic attacks is very well possible for a CPAS.

In this chapter, we will outline how a CPAS can be secured today using common security means, as well as the security features built into a modern CPAS product.

2.7 IT Security for Automation Systems

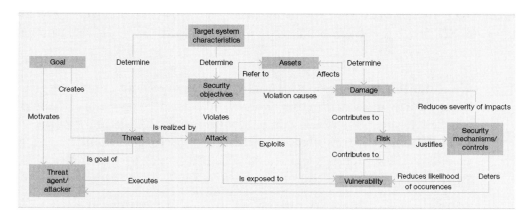

Figure 2–39 Security Threat Analysis Terms and Their Interrelations (*ABB Review*, 2/2005)

IT Security Risk Management for CPAS

Any security solution should be the result of a thorough risk analysis to ensure that the budget available for security is spent effectively and efficiently on staffing and mechanisms to bring the residual risk (risk after applying measures) down to a level that is acceptable to, and consciously accepted by, the plant owner. Figure 2–39 shows the various facets that need to be considered during such a risk analysis.

In this section, we will define assets and security objectives because these concepts are needed to later describe the rationales for, and effect of, the security measures implemented in a CPAS. A more detailed treatment of risk analysis for automation systems and underlying concepts can be found in VDI/VDE 2182 (2007) and ANSI/ISA-99.00.01-2007.

Security measures always have the goal of preventing something bad from happening to a valuable good. In IT security terminology, a valuable good that can incur a direct or indirect loss is called an *asset*. This may be production data, application programs, humans in the plant, and also intellectual property such as product recipes or even the reputation of the company. The "something bad" that can happen to an asset is expressed as a violation of certain properties, which are the security objectives for the asset.

Security objectives thus describe what fundamental types of threats the system is to be secured against. The three security objectives explained in the following offer a framework for categorizing and comparing the security mechanisms of various systems. (Note that in the discussion of system dependability in Section 2.5, the terms integrity and availability have different meanings than they do here.)

Confidentiality—The confidentiality objective refers to preventing disclosure of information to unauthorized persons or systems. For automation systems, this is relevant both with respect to domain-specific information, such as product recipes or plant performance and planning data, and to the secrets specific to the security mechanisms themselves, such as passwords and encryption keys.

Integrity—The integrity objective refers to preventing undetected modification of information by unauthorized persons or systems. For automation systems, this applies to information such as product recipes, sensor values, or control commands. This objective includes defense against information modification via message injection, message replay, and message delay on the network. Violation of integrity may cause safety issues, that is, equipment or people may be harmed.

Availability—Availability refers to ensuring that unauthorized persons or systems cannot deny access or use to authorized users. For automation systems, this refers to all the IT elements of the plant, including control systems, safety systems, operator workstations, engineering workstations, and manufacturing execution systems, as well as the communication systems between these elements and to the outside world. Violation of availability, also known as Denial-of-Service (DoS), may not only cause economic damage but may also affect safety as operators may lose the ability to monitor and control the process.

Some of these security objectives are to a certain extent independent of each other, and for most systems only a subset of the security objectives will have a high priority or be applicable at all. In contrast to many enterprise and government information processing systems, the priority ranking of the security objectives is different in control systems. Availability, both of the production facilities themselves as well as the ability to be in control of the plant, has highest priority, followed closely by integrity of data and commands. Confidentiality has typically a much lower priority.

Principles of Secure System Design

There exist a couple of timeless security design principles, which apply to any system, including automation systems. While a thorough treatment would lead too far in this chapter—such treatment can be found in Anderson (2008) and other books on security engineering—it is worthwhile to quickly review a few of those principles, which in the past have often been disregarded in securing industrial automation systems.

Any (security) system is only as secure as its weakest element. An attacker will actively try to find the weakest element; therefore, for a given budget, one should strive to achieve a balanced defensive strength across all system components and attack vectors.

Security has to be maintained during the entire plant lifetime. Due to the discovery of new attacks and vulnerabilities, as well as the availability to attackers of more powerful computers, any technical security architecture will begin to lose effectiveness from the moment of deployment. In addition, misconfigurations of technical mechanisms and other process and procedure failures can easily weaken any security measure. A continuous effort from a technical and procedural perspective is thus necessary to maintain the desired security level (Brändle and Naedele, 2008).

As any security mechanism may be overcome by an attacker, it is advisable to design the system architecture in a way that the most sensitive parts of the system are protected by multiple rings of defense that all have to be breached by the attacker to get to the "crown jewels." Such an architecture, like the wall systems of a medieval castle, is called "defense in depth." In addition, not only protection mechanisms should be deployed, but also means of detecting attacks. This includes both technical measures, such as intrusion detection systems, as well as procedural measures, such as review of log files or access rights.

No user should be able to do more in the system than what he or she needs for the job. This "principle of least privilege" limits the consequences of both insider attacks and externally compromised user accounts.

For a long time, owners of industrial plants relied for security on the fact that automation systems and their protocols were not very well known in the IT community. From the fact that other applications, like operating systems, are much more frequent known targets of attacks, they assumed that automation systems are secure. However, this assumption of "security by obscurity" has been shown to be wrong (Byres 2002). Sufficient information on automation systems is easily available to any interested party, and there are a significant number of malicious interested parties, such as insiders, organized crime, issue activists, or national security services.

CPAS Security Scenarios

A typical misconception when looking at securing an automation control system is that attacks can only originate from company-external networks (e.g., from the Internet) and can only reach the control system if the control system is directly connected to the company-external network. While this is of course a valid scenario, it is far from complete. This section looks at three common security scenarios for CPAS and outlines how security measures can be arranged in multiple lines of defense.

Defending Against Attacks from External Networks / Internet

Today's automation control systems often have connections to system-external networks. System-external networks not only include the Internet but also office networks or in general any network that is not part of the control network itself. The first and most important step in defending against attacks from such external networks is to have a true understanding of all such connections. Once such an overview exists, several steps can and should be taken to secure the control system against attacks from system-external networks.

>   **Network separation–**The control system should be clearly separated from any external network. This can be achieved by, for example, using firewalls to control data access to the control network. In order to authenticate accessing entities, the combination of a firewall with a Virtual Private Network (VPN) gateway is a good solution. A more secure architecture is to work with a so-called demilitarized zone (DMZ), a zone that serves as a proxy between external networks and the control system.
>
>   **Restricted data flows–**Data flows into and out of the control system must be strictly controlled and limited to the absolutely necessary. If possible, the direction of data flows should be from the inside to the outside (i.e., data flows should be initiated by the control system). If

2.7 IT Security for Automation Systems

there is a need to exchange data with external network, then this should be done using an intermediary system within the DMZ.

Terminal servers–Securing remote, interactive access to the control system can be best achieved if direct access is not allowed, but the remote entity has to go through a terminal server that resides within the DMZ. The remote client is required to connect to the terminal server first (e.g., using a desktop mirroring application) and can then from there access the control system. The terminal server should not have any specific control system functionality or application and should be operated with the strictest security settings (i.e., it should have anti-virus and host intrusion detection software running and it must be patched regularly).

Protected communications–Communications both within the control system but especially with external networks should be protected using encryption and possibly message integrity protection. Because many of the currently used control system protocols do not inherently support such mechanisms, they can be tunneled through VPNs.

Application separation–Applications that are not relevant for control system operations, such as email clients or Web browsers, should not be installed on any computers that are part of the control system. If such applications are needed, then they should run on separate machines that reside in a dedicated network zone that is separated from the control system.

Monitoring & audit–It is not only important to have mechanisms in place to protect the control system against attacks from external networks, but also that these protection mechanisms are monitored and audited regularly. This includes reviewing of log files or auditing security settings. Irregular entries in log files or changed security settings can be a sign of an attempted or successful attack.

Containment in CPAS–The mechanisms mentioned so far can be thought of as the first lines of defense, with the purpose of shielding the control system from external attacks. However, there must not only be layers of defense in place around the control system but also mechanisms within the control system itself.

The first step should be to harden every single host within the control system and thus minimize the surface of attack. Hardening includes restricting applications and open ports and services to an absolute minimum. It is not enough to stop unneeded applications; they should be completely uninstalled so that an attacker cannot start them. System hardening must also look at user accounts and ensure that only needed accounts are installed (e.g., no guest accounts) and that strong authentication is enforced.

Hosts within the control system should further be secured by having up-to-date patches installed and anti-virus or host intrusion detection software running. Installing patches without prior testing must not be done because the patches might have a negative effect on the control system. Figure 2-40 shows how a CPAS vendor can help by providing information based on the testing of patches on reference systems.

The network security of a CPAS can be improved by using intrusion detection systems to monitor traffic flows.

Defending Against Malware from Service/Engineering Laptops and Portable Media (e.g., USB Sticks)

Besides static, direct connections between the control network and external networks there also exist temporary, indirect connections that are often not considered when securing a control system. Such temporary, indirect connections are, for instance, mobile devices such as service laptops that are connected to the control network by an engineer, or portable media such as USB sticks or CDs that are connected to computers within the control network.

In an optimal scenario both mobile devices and portable media that are used within the control network should only be used within the control network and nowhere else. However, in reality this is rarely the case. The control system must therefore be protected from any attacks; for example, the spreading of a virus, which could originate from a mobile device or portable media.

Protecting against attacks originating from mobile devices can be done by only allowing the connecting of such devices at dedicated points within dedicated zones in the control systems. These zones should be separated from the control network with a firewall and should contain intrusion detection systems and malware scanners. The dedicated zone should also create detailed audit logs of all activities. This is important for doing forensic analysis in case of a security incident and to trace the origin of the incident.

2.7   IT Security for Automation Systems   125

| Microsoft | | | | ABB | |
Security Bulletin	KB Articles	Severity	OS or Product	Validation Status	Comments
September 2008					
MS08-055	955047 953405 953404 951944	Critical	Office XP Office 2003 Office 2007	Q	
MS08-054	954154	Critical	Windows Media Player	N/A	
MS08-053	954156	Critical	Windows Media Encoder	N/A	
MS08-052	954593 938464	Critical	GDI+ Windows XP Windows Server 2003 Windows Vista Windows Server 2008 E6.0 SP1 on Windows 2000	Q	
	947739 947742 947746 947748 953405 954478 954326 954606		.NET Framework 1.0 .NET Framework 1.1 .NET Framework 2.0 .NET Framework 2.0 SP1 Office XP Office 2003 Office 2007 SQL Server 2005		

Figure 2–40   Extract from Document Showing the Applicability and Status of Released Microsoft Patches for a CPAS

The use of portable media should be restricted as much as possible. If such use is necessary then such media should first be checked for malware before being connected to a control system host. This can be done with a dedicated "Malware-Scanning Station." This dedicated computer should not be connected in any way to the control network and should have up-to-date malware detection software running.

Defending Against Insiders

Quite a few of the IT security incidents that have happened in industrial plants were actually not conducted by an attacker via the Internet, but by insiders. As insiders already have (partial) legitimate access to the system as well as detailed knowledge, they often make the most dangerous attackers.

Defending against insiders relies on two main approaches:

- Turning the insider into an "outsider" for most of the system by introducing access control that limits each user to the minimum access to get the job done ("least privilege").

- Deterring insiders from turning bad by creating personal accountability and enforcing punishment for acting against the policy.

The first line of defense against inside attacks is prevention of unauthorized access. Modern applications that encapsulate security-critical data or commands, like database management systems or operating systems, allow defining in detail what each user is allowed to do for each critical item (or asset) in the system, such as a database table or field, or a file in the file system. A modern CPAS provides the same features, for example, by assigning Security Definition Aspects to objects in the automation system (see Figure 2–41).

In order for a user to be able to execute an action in the CPAS, he or she must possess the specific permission that has been defined as "required" for the desired action and target object. Example permission types in a CPAS are Read, Configure, Operate, Tune, Shutdown, Security Configure, etc. The security administrator of the system may create additional permission types.

In a CPAS with thousands of objects and hundreds of named users, the administration of access rights becomes a rather burdensome task. This will result either in high costs for the security maintenance, or the administration staff will take shortcuts like assigning all users the same rights, which will severely reduce the effectiveness of the security measures. A CPAS should support two mechanisms to help in keeping fine-grained security definitions manageable:

- Definition of access control statements on automation objects according to different hierarchies and types, and flexible inheritance/filtering rules, so that, for example, the access rules can be assigned to multiple objects, which might include all objects in a certain plant area or all objects with a certain function (such as valves or control code), with a few configuration actions, avoiding repetitive data entry.

- Role-based access control (RBAC) instead of user-based access control. In any plant there will be many users whose access rights are the same based on their job description (e.g., plant operator, maintenance technician, and shift supervisor). By managing all those users in terms of groups according to their roles instead of as individuals, permissions need to be carefully designed only for a few roles instead of for hundreds of individuals (see Figure 2–42). Individual user accounts then only need to be touched in terms of role/group assignments whenever a user joins, leaves, or otherwise changes his or her job description. Of course, for fast engineering the system should be flexible enough to allow the plant-specific definition of group/role types in addition to some predefined common role types.

2.7 IT Security for Automation Systems

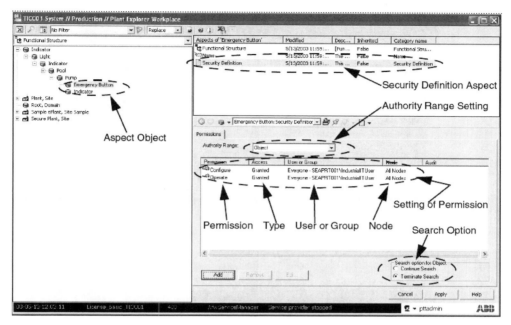

Figure 2–41 Definition of Access Rights for Automation Objects

When defining access control for a CPAS, it should also be considered whether certain activities should be restricted according to the location of the user (e.g., only at certain workstations). Some plants have the policy that certain safety-critical activities can only be initiated from consoles within the plant, not via remote access.

Another access limitation consideration is to modify the user interface based on the role of the user (e.g., by hiding or disabling certain control elements) (see Figure 2–43). This has several benefits. By reducing the choices available to the user to those that are needed for the role, training effort and cost are lowered and the likelihood of unintentional mis-operation is also reduced.

In addition, certain user interface limitations, like disabling access to the underlying operating system or to applications like web browsers, prevent the user from circumventing the CPAS-specific access controls. A further step would then be to disable not just access to non-CPAS applications on the CPAS workstations, but (as we have seen) also to use technical measures like VPNs to prevent additional computers (e.g., unauthorized laptops) from connecting to the CPAS network and hosts.

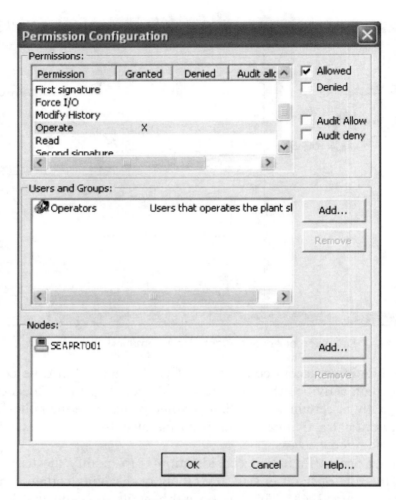

Figure 2–42 User Interface for Role-based Access Right Definition

As mentioned earlier, the second line of defense against insiders is deterrence. This means that there should be a high likelihood that any unauthorized activity will be detected and that the responsible party will be held responsible. Of course, a successful deterrence strategy requires not just technical means, but also security policies and work contracts that allow the company to actually punish any offender.

The technical means mentioned in this context are necessary to detect that an unauthorized or malicious activity has taken place and to identify who did it. The evidence may have to stand up to legal scrutiny.

2.7 IT Security for Automation Systems

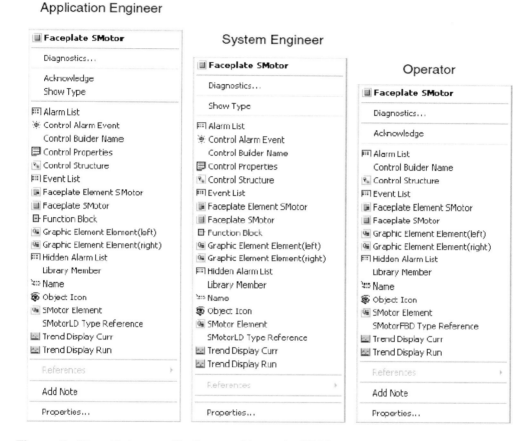

Figure 2–43 Role-specific Context Menus in CPAS

For detection, one can use CPAS-external tools like file system integrity checkers, network intrusion detection systems, virus scanners etc., or operating system level and CPAS internal audit logs (see Figure 2–44).

While the timely detection of certain unauthorized activities may have some value in order to limit their consequences on plant operation, deterrence requires that the activity can be undeniably tied to an individual user. This precludes the use of shared accounts or shared credentials (passwords) and requires the audit functionality itself and the storage of the audit logs be to a high degree tamper-proof. Some domain regulations like FDA CFR11 require digital signatures of the operator for every relevant action. Read, Modify, and Delete access to audit logs should be highly restricted.

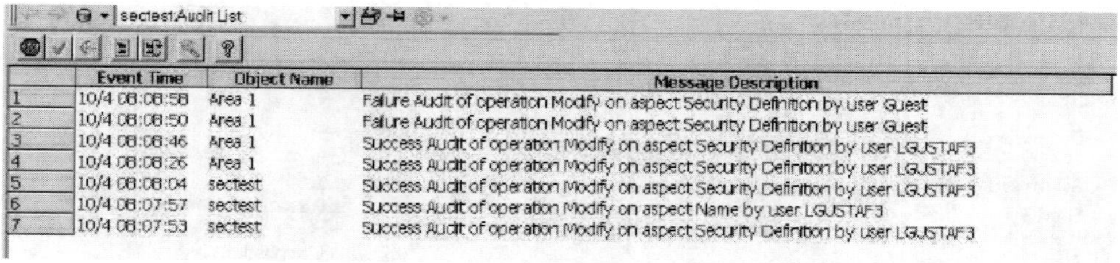

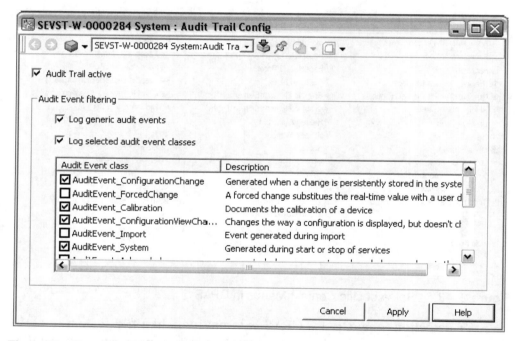

Figure 2–44 Audit Log in CPAS

Figure 2–45 Audit Trail Configuration Dialog

The CPAS should provide comprehensive audit facilities, including the ability to log successes and failures of a large variety of system commands (see Figure 2–45). To reduce security management effort, audit trail configuration should be integrated with access control configuration.

In order to verify the security configuration settings, especially if there are non-trivial hierarchical dependencies, it is helpful if the CPAS offers a function to list the current security configuration or even to simulate the effects of a change before it is actually applied (see Figure 2–46).

2.7 IT Security for Automation Systems

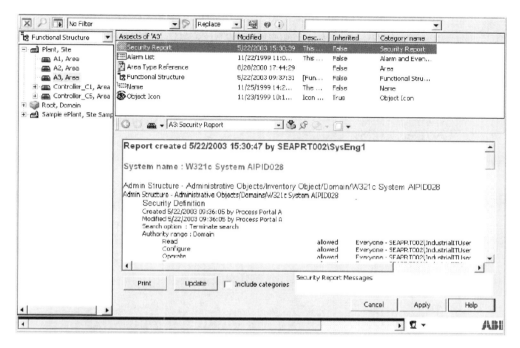

Figure 2–46 Security Configuration Report

Conclusion

This chapter looked at IT security in the context of a CPAS. IT security is certainly a new and important challenge for industrial control systems, especially if they have extensive communication connections inside and outside the plant and enterprise. However, with commonly available technical security means, generally accepted IT security policies and practices, and the specific security-related features in modern CPAS applications, such systems can be deployed and operated very securely so that security concerns should not prevent anybody from reaping the benefits of a CPAS.

References

Anderson, R.J. *Security Engineering: A Guide to Building Dependable Distributed System*. Hoboken: John Wiley & Sons, 2008. ISBN: 978-0-470-06852-6.

American National Standards Institute/International Society of Automation, *ANSI/ISA-99.00.01-2007, Security for Industrial Automation and Control Systems: Concepts, Terminology and Models.* Research Triangle Park: ISA, 2007.

Brändle, M. and Naedele, M. "Security for Process Control Systems: An Overview." *IEEE Security & Privacy* Vol. 6, No. 6, pp. 24–29 (Nov./Dec. 2008).

Byres, E. J. "The Myth of Obscurity." *InTech*, p. 76, Research Triangle Park: ISA, Sep. 2002.

VDI/VDE 2182, VDI Guideline IT Security for Industrial Automation— General Model, Sheet 1; Draft for Comments." Gründruck VDI, Aug. 2007. Beuth-Verlag, Berlin, 2007.

2.8 Safety—One of the Key Practices in Control Engineering

Zaijun Hu

Introduction

The current developments in automation and control engineering show a strong trend that safety is an essential and inevitable discipline in system development, production, engineering and service. The demand by customers for conformity to the related safety standards, such as IEC 61508, IEC 61511, IEC 62061 and ISO 13849, forces device manufacturers and system providers to implement the required safety features.

In an increasingly multidisciplinary engineering environment, and in the face of ever-increasing system complexity, there is a growing need for all engineers and technicians involved in process engineering to be aware of the implications of designing and operating safety-related systems. This includes knowledge of the relevant safety standards, safety development, safety engineering, and safety service, for which a solid understanding and flexible use of the engineering approaches for systems including hardware and software are indispensable preconditions. The chapter will highlight some of the key aspects of safety-related development and engineering practices, including the basic definitions, the related safety standards, technical challenges and constraints, typical system architecture, and implementation approaches.

Safety, Functional Safety and Four Implementation Components

Based on IEC 61508, **safety** is defined as freedom from unacceptable risk of physical injury or of damage to the health of people, either directly or indirectly, as a result of damage to property or to the environment (IEC 61508-4,

1998). *Functional safety* is part of overall safety relating to the equipment/process under control and the corresponding control system, which depends on the correct functioning of the safety-related system.

For example, an over-temperature protection device, using a thermal sensor in the windings of an electric motor to de-energize the motor before it can overheat, is an instance of functional safety. But providing specialized insulation to withstand high temperatures is not an instance of functional safety (although it is still an instance of safety and could protect against exactly the same hazard).

A *safety function* is intended to achieve or maintain the safe state of the process or equipment under control, in respect to a specific hazardous event. The safe state is a state of process or equipment operation where the hazardous event cannot occur. Absolute safety, where risk is completely eliminated, can never be achieved with acceptable effort and cost; risk can only be reduced to an acceptable level.

Functional safety based on IEC 61508 uses *safety integrity level* (SIL) to measure the intended reliability of a system or function. Four levels are defined, with SIL 4 providing the highest reliability and SIL1 the lowest. An SIL is determined based on a number of qualitative and quantitative measures. A safety-related system which implements functional safety and provides conformity to the related safety standards is supposed to deliver evidence and convincing arguments to the responsible authorities/certification bodies based on such measures. Examples of qualitative measures are:

- Program temporal and sequence monitoring
- Plausibility and boundary value check
- Supervision of variable and invariable memory
- Data flow monitoring

The quantitative measures include PFD (probability of failure to perform a safety function) and PFH (probability of failure per hour).

Implementation of functional safety needs *four key components*: management of functional safety, product safety measures in the system, a safety case, and competence. Figure 2–47 illustrates the structure.

> **Safety measures in the system**—A safety-related system needs to implement the required safety measures, which are needed to reduce

2.8 Safety—One of the Key Practices in Control Engineering

Table 2–2 Safety Integrity Level vs. Quantitative Measures

SIL	Average probability of failure to perform a safety function on demand (Low demand mode of operation)	Probability of a dangerous failure per hour (High demand or continuous mode of operation)
4	$\geq 10^{-5}$ to 10^{-4}	$\geq 10^{-9}$ to 10^{-8}
3	$\geq 10^{-4}$ to 10^{-3}	$\geq 10^{-8}$ to 10^{-7}
2	$\geq 10^{-3}$ to 10^{-2}	$\geq 10^{-7}$ to 10^{-6}
1	$\geq 10^{-2}$ to 10^{-1}	$\geq 10^{-6}$ to 10^{-5}

Figure 2–47   Key Components in the Realization of Functional Safety

dangerous risks and achieve the targeted safety integrity level. An appropriate architectural approach is needed to address challenges such as failure detection, isolation, and reaction. In addition, some architectural constraints such as real-time requirement, concurrency, availability, reliability and balance between system cost and engineering cost are essential technical and economic factors for a safety-related system.

Management of functional safety—A safety-related system should be reliable. Its failures should be dealt with based on the required safety integrity level. Two failure categories are introduced in functional safety: random hardware failure and systematic failure. *Random hardware failure* occurs at a random time, and results from one or more of the possible degradation mechanisms in the hardware. *Systematic* *failure* is related in a deterministic way to certain causes in specification, design, implementation, manufacturing, configuration, commissioning, operation, maintenance and support. The key issue of functional safety management is to deal with systematic failures in different phases of a product's life cycle such as configuration, installation, commissioning and operation. Verifiable and controllable management, including a consistent development and engineering process with the appropriate approaches for specification, design, implementation, verification, validation and documentation, is crucial in functional safety implementation. A SIL-compliant engineering tool is an efficient measure to avoid failures in engineering.

Safety case—A safety case is a set of related claimable evidence for the arguments that support fulfillment of a safety-related technical or management requirement in a manner capable of convincing a safety assessor. Identification, collection and documentation are required to create a safety case. A suitable documentation technique is necessary.

Competence—The efficacy of the measures will be severely hindered if the people responsible for their implementation do not have the required competence. Implementation of functional safety is a multi-disciplinary practice, which needs not only comprehensive technical competence but also substantial management skills. Technical competence covers knowledge of the related safety standards, experience and competence in hardware and software engineering, capability in concept and architecture design, and systematic and structured documentation techniques, while management skills address the systematic supervision and control of the required safety-related activities, as well as the perception and handling of human factors.

Safety Instrumented Systems

Safety instrumented systems (SIS) are an implementation form of functional safety for industrial processes such as oil refining, power generation and chemical reactions. A SIS is formally defined in IEC 61508 as "a distinct, reliable system used to safeguard a process to prevent a catastrophic release of toxic, flammable, or explosive chemicals." A SIS performs specified safety instrumented functions (SIF) to achieve or maintain a safe state of a process when unacceptable or dangerous process conditions are detected. Typical safety instrumented systems are:

- Emergency shut-down and process shut-down systems
- Burner management and boiler protection
- Interlock systems
- High-pressure protection systems
- Pipeline protection systems
- Fire and gas systems

The correct operation of a SIS requires a series of pieces of equipment to function properly. It must have sensors capable of detecting abnormal operating conditions, such as high flow, low level, or incorrect valve positioning. An input interface connects a sensor with a logic solver, which processes the signal from the sensor and changes its outputs according to user-defined logic. The logic solver may use electrical, electronic or programmable electronic equipment, such as a micro-controller, PLD (programmable logic device), or ASIC (application-specific integrated circuit). The signal from the logic solver will act on the final element, such as a valve or contactor, via an output interface. The related support system for configuration and parameterization should be designed to provide the required integrity and reliability.

Architectural Constraints

A safety-related system should be able to address a set of architectural challenges, which need to be taken into account in the design and engineering of a safety instrumented system.

> **Availability**—Availability is a measure that defines the likelihood that a process is operable. The availability of a process control system can be increased through the parallel use of system components or redundancy. In many cases, availability is contrary to safety. A safety instrumented system is intended to achieve or maintain the safe state of a process, which typically means that the related processes is required to be stopped, and thus not operable or available any more. It is therefore

important in safety development and engineering to take availability into account. A good safety instrumented system should be able to provide a well-balanced weighting of availability and safety.

Integration—Integration means that the same system hardware, the same engineering tool, and the same HMI (human-machine interface) are used for Basic Process Control Systems (BPCS) and SIS. Traditionally, BPCS and SIS are separated from each other. The recent trend shows that more and more systems where they are integrated together are delivered. That means they use the same hardware, HMI, and engineering tool. This will reduce the system hardware cost, the effort for engineering, configuration, commissioning, and operation; however, in an integrated solution it is crucial to ensure the logical independence of the safety-related parts from the non-safety related parts.

Logical independence—Logical independence means that failures in the non-safety related parts should not lead to failures of the related SIS so that the SIS cannot fulfill its function of protecting the process under control from hazards. Logical independence can be realized through resource and time partitions (i.e., it must be ensured that a safety instrumented system will get the required resources and time to perform its safety instrumented functions). Logical independence can also be achieved if all failures in the non-safety-related parts that are related to the performance of the safety instrumented functions are directly or indirectly detected and controlled accordingly.

Safety integrity measures—Safety integrity measures are those that must be implemented to achieve the targeted safety integrity level. IEC 61508 specifies a set of safety integrity measures with a rating for each measure. Table 2–3 gives some examples of those measures.

It is necessary to take those measures into consideration on the architectural level in order to clarify any structural constraints that may have an overall impact on the whole system implementation.

Modularity—Modularity is a way to manage the complexity of a system and also provide the needed flexibility for safety applications. IEC 61508 proposes the use of subsystems to construct the system architecture. In addition to the normal principles of modularity, such as clearly defined functions of the subsystems with interfaces that are as simple as

2.8 Safety—One of the Key Practices in Control Engineering

Table 2–3 Safety Integrity Measures

Safety integrity measures	SIL 1	SIL 2	SIL 3	SIL 4
Fault detection and diagnosis	---	R	HR	HR
Error detecting and correcting codes	R	R	R	HR
Safety bag techniques	---	R	R	R
Recovery block	R	R	R	R
Backward recovery	R	R	R	R
Re-try fault recovery mechanisms	R	R	R	HR
Program sequence monitoring	HR	HR	HR	HR

Note:

HR The technique or measure is highly recommended for this safety integrity level. If this technique or measure is not used then the rationale behind not using it should be detailed during the safety planning and agreed upon with the assessor.

R   The technique or measure is recommended for this safety integrity level as a lower recommendation to an HR recommendation.

--- The technique or measure has no recommendation for or against being used.

possible, it is also necessary to consider reusability and the related safety cases associated with each subsystem. SIL-compliant subsystems, which not only include logic solvers but also field instruments, I/O modules, and actuators, help to reduce the effort require to construct an entire safety loop.

A complete supply of all kinds of SIL-compliant subsystems is crucial for successfully addressing the safety requirements of a plant. Modularity offers many interconnection options, thus making these subsystems suitable for all safety and critical process automation applications. Advanced diagnostics and built-in checks, which are required to achieve the SIL compliance of subsystems, will prevent the inadvertent degradation of safety applications.

Management and engineering of functional safety—To address the variety of requirements a safety instrumented system needs flexibility and adaptability, which can be accomplished by a SIL-compliant system engineering environment. Additionally, the safety engineering environment should help to efficiently address the requirements of functional safety management, which covers the entire safety system life cycle from planning, through design and library management, to commissioning and support. A set of measures to avoid systematic faults should be implemented in the engineering system. Safety concepts and measures, such as safeguard for preventing systematic faults due to non-SIL compliant configuration, are necessary. Once identified as a safety application, the engineering system will automatically limit user configuration choices and will prevent download if SIL requirements are not met. In addition to SIL-compliant engineering capability for the configuration of safety applications, the engineering system should also provide SIL-compliant standard and application-specific libraries.

Functional safety standards—Functional safety standards have a huge influence on the design of the system architecture of a safety instrumented system. Safety standards can be categorized into basic safety standards and domain-specific safety standards. Table 2–4 provides an overview of important safety standards that may need to be taken into account.

Typical Components of a Safety Instrumented System

Figure 2–48 illustrates a typical system architecture of a SIS, which includes a instrumented system for protecting a process under control from hazards, an engineering system, and a basic process control system (BPCS) for standard process control.

Traditionally, the BPCS has been physically separated from the SIS to ensure the independence of safety-related parts from non-safety related parts. This solution normally requires high system cost and more system engineering and maintenance effort.

In addition to the regular engineering functions, the engineering system should also provide features to support the management of functional safety to avoid systematic faults, which, as has been mentioned, can occur in safety application engineering, including specification, planning, design, library management, commissioning, and support. It should show capability in the management of safety-critical data, configuration, and delivery of the required safety cases.

2.8 Safety—One of the Key Practices in Control Engineering

Table 2–4 Safety Standards

Safety Standards	Description
IEC 61508	Functional safety of electrical/electronic/programmable electronic safety-related systems
IEC 61511	Functional safety—Safety Instrumented Systems for the Process Industry Sector
IEC 62061	Safety of machinery—Functional safety of safety-related electrical, electronic and programmable electronic control systems
ISO 13849	Safety of machinery—Safety-related part of control systems
IEC 61513	Nuclear power plants—Instrumentation and control for systems important to safety

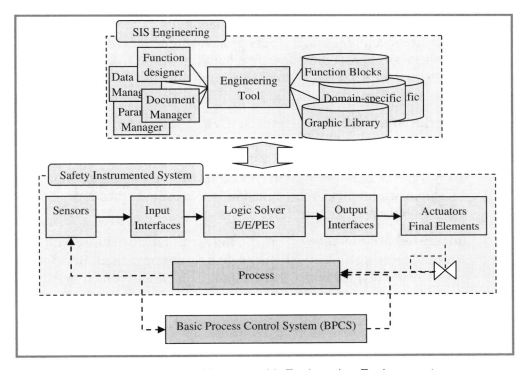

Figure 2–48 SIS System Architecture with Engineering Environment

Architecture Alternatives

Safety instrumented systems may have three different architectures: separate, interfaced, and integrated.[3] Each one has its advantages and drawbacks, which should be contemplated in the design of a safety instrumented system. Issues with system cost, common cause error, and level of effort for engineering, learning, training, and maintenance are the crucial factors in choosing the system architecture.

> **Separate architecture**—The basic process control system and the safety instrumented system are from two different suppliers and have two different engineering systems. This architecture was once the preferred solution for many applications that needed to fulfill safety requirements. Due to different vendors and different engineering systems and operation stations, the architecture normally has high system cost and considerable engineering, learning, training, and maintenance effort, resulting in high operational expenditures. The primary benefit of this architecture is that it is less prone to common-cause errors. An interface is used for communication between the basic process control system and the safety instrumented system. Figure 2–49 shows the separate architecture.
>
> **Interfaced architecture**—Both BPCS and SIS are provided by the same supplier; however, they are not interchangeable, although they look very similar. There are two separate engineering tools, respectively, for BPCS and SIS. Both BPCS and SIS may share the same development station and the same kinds of wiring. Figure 2–50 illustrates an interfaced architecture.
>
> **Integrated architecture**—This is a fully integrated solution, with the same engineering tool, development, and operation station. Both BPCS and SIS are provided by the same supplier and implemented in the same physical unit. Figure 2–51 illustrates this architecture.

Table 2–5 shows a comparison between these three architecture alternatives.

---

3. There are different interpretations of separate, interfaced and integrated. Here we show one of them.

2.8 Safety—One of the Key Practices in Control Engineering

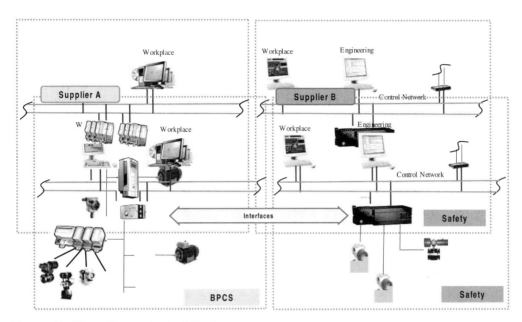

Figure 2–49 Separate Architecture

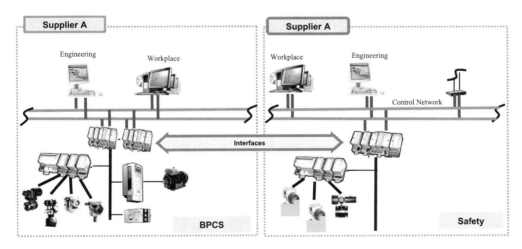

Figure 2–50 Interfaced Architecture

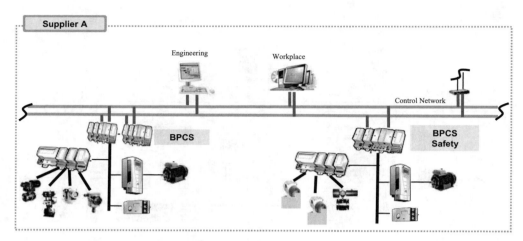

Figure 2–51   Integrated Architecture

Table 2–5   Architecture Alternatives

	Separated	**Interfaced**	**Integrated**
Characteristics	Historically appreciated for near total avoidance of common mode failures. Separate and different BPCS & safety systems	Risk of common mode failures reduced and managed through internal diversity, software diagnostics	Common cause error may be high
Engineering	Separate engineering tools	Similar but separate engineering tools	Same engineering tool
Operation/ Maintenance	Separate/dual systems and HMIs	Separated systems	Same system and HMI
Training/ Learning	High	Reduced	Low
Life-cycle management	Different	Possibly Common	Common

Conclusion

Functional safety is not only a feature of a system but also a capability of the related development process that is applied to develop a system with the safety requirements. The management of functional safety is crucial to avoid systematic faults in a system. For the development of a safety-related system, it is important to take the related architecture constraints into account. Availability, integration, flexibility, adaptability, independency, safety integrity measures, and the related safety standards are some factors that need to be considered in the development. Separate, interfaced, and integrated architectures are well-known design alternatives.

References

International Electrotechnical Commission, *IEC 61508-1 Functional safety of electrical/electronic/programmable electronic safety-related systems—General requirements.* Geneva: IEC 1998.

Bibliography

International Electrotechnical Commission, *IEC 61508-2 Functional safety of electrical/electronic/programmable electronic safety-related systems—Requirements for electrical/electronic/programmable electronic safety-related systems.* Geneva: IEC 2000.

——— *IEC 61508-3 Functional safety of electrical/electronic/programmable electronic safety-related systems—Software requirements.* Geneva: IEC 1998.

——— *IEC 61511 Functional safety—Safety instrumented systems for the process industry sector.* Geneva: IEC 2003.

——— *IEC 62061 Safety of machinery—Functional safety of safety-related electrical, electronic and programmable electronic control systems.* Geneva: IEC 2005.

——— *IEC 61513 Nuclear power plants—Instrumentation and control for systems important to safety.* Geneva: IEC 2001.

International Organization for Standardization *ISO 13849 Safety of machinery—Safety-related parts of control systems.* Geneva: ISO 2006.

Börcsök, J. "Elektronische Sicherheitssysteme: Hardwarekonzepte, Modelle und Berechnung." Hüthig Verlag Heidelberg, 2007.

Smith, D. and Simpson, K. "Functional Safety, Second Edition: A Straightforward Guide to IEC 61508 and Related Standards." Oxford: Butterworth-Heinemann, 2004.

Wratil, P. and Kieviet, M. "Sicherheitstechnik für Komponenten und Systeme Sichere Automatisierung für Maschinen und Anlagen." Hüthig Verlag Heidelberg, 2007.

2.9 Traceability

Martin Hollender

Introduction

Traceability is defined as the ability to identify and analyze all relevant stages that lead to a particular point in a process. For sensitive products like pharmaceuticals or food, traceability is required by FDA 21 CFR Part 11 to achieve accountability. For example, if milk powder does not contain the correct mix of ingredients, babies can die. It is therefore very important to be able to confirm in detail, at every step, how the milk powder was produced and what ingredients were used. Also, for industries involving risks for people or the environment such as power plants, refineries or chemical plants, complete traceability is more and more required to be able to reconstruct accidents and incidents. In addition, an audit trail helps to maintain data integrity with respect to changes. Another benefit of traceability is the capability of understanding the history of plant production as a basis for the optimization of future production. Even if traceability as defined by FDA 21 CFR Part 11 is not a legal requirement for all manufacturers, many parts provide a good model for traceability for any manufacturer or distributor even outside the food and drug industries.

FDA 21 CFR Part 11—Title 21 Code of Federal Regulations

The most important regulation concerning traceability is FDA 21 CFR Part 11, issued by the U.S. Food and Drug Administration (FDA). This rule applies to all FDA-regulated products manufactured in the United States, and also to FDA-regulated products manufactured elsewhere but distributed in the United States, which gives it international relevance. FDA 21 CFR Part 11 has two main areas of enforcement, electronic records and electronic signatures, and includes the following requirements:

- Validate systems to ensure accuracy, reliability, consistency, intended performance, and the ability to discern invalid or altered records. This includes validating commercially available software.
- Maintain the ability to generate accurate and complete records in both human-readable (records that are read by a person as opposed to a machine) and electronic form so that FDA may inspect, review, and copy the records.
- Protect records so that they are readily retrievable throughout the required retention period.
- Limit system access to authorized individuals.
- Record changes in a manner that does not obscure previous entries.
- Use secure, computer-generated, time-stamped audit trails to independently record the date and time of operator entries and actions. The audit trail must also be available for FDA review and copying and must be retained for at least the same period as the records.
- Use operational system checks to enforce sequencing steps.
- Use authority checks to ensure that only authorized individuals can use the system, electronically sign a record, access the operation or computer system, alter a record, or perform the operation at hand.
- Use device checks to determine the validity of data input or operational instructions.
- Ensure that authorized users have the appropriate education, training, and experience.
- Establish and follow written policies that deter record and signature falsification.
- Establish adequate controls over access to all system documentation as well as over the distribution and use of such documentation. Most important, establish controls over highly sensitive documentation, such as instructions on how to modify system security features.

2.9 Traceability

- Establish adequate controls over revisions and changes, and maintain audit trails of modifications to system documents.

As an additional benefit, moving to a paperless world with fully electronic data handling promises cost savings from improved efficiency and reduced physical handling and storage.

Electronic Records

FDA 21 CFR Part 11 defines an electronic record as any combination of text, graphics, data, audio, pictorial, or other information representation in digital form that is created, modified, maintained, archived, retrieved, or distributed by a computer system. The regulation's controls are designed to ensure the authenticity, integrity, and confidentiality of electronic records and minimize the possibility of easy or inadvertent repudiation of the electronic record by the signer.

Electronic Signatures

Electronic signatures (Figure 2–52) are given legal equivalence with traditional "wet ink" signatures on paper. Critical operations in a process or critical changes of machine settings might require significant authentication steps to ensure that the correct persons have the rights and authority to perform certain actions. The user will be required to provide username, password and new value and, if necessary, comments or a supervisor signature before the change affects the process.

CPAS Support for Traceability

A CPAS must facilitate compliance with the rules of FDA with features like system security, secure data management and reporting, electronic records and signatures, and the automated electronic recording of changes. Core system functions that support regulatory requirements are:

- Access control
- Authentication and re-authentication
- Alarm and event management

Figure 2–52 Authentication Dialog for Digital Signature

- Audit trail
- Batch control and management
- Electronic recording
- Electronic signatures
- Event reports
- History and archiving
- Trends
- Redundancy
- Sequential function charts
- Shift, production and batch reports
- Infrastructure and network security

For login, authentication and electronic signing, two component security codes should apply. Every combination of user identification and password needs to be unique. Minimum password length, password aging, and rules against re-using recent passwords must be configurable. Access should be controllable from the system level down to the object level (e.g., a single valve or a

2.9 Traceability 151

range of cleaning equipment or an entire ingredient list). Access to functional inputs must be limited, including the right to open a single valve, or start a process, or schedule the next batches and campaigns. Operators must identify themselves both at login and before an input is accepted; for example, before a motor is switched on or a cleaning process is started.

Change Control

In addition to authority checks, a CPAS should give manufacturers control over which changes are made. Using electronic signatures, system users can be held accountable and responsible for actions initiated. This could apply, for example, to the configuration of control applications, including all embedded logic and I/O functions for release to production.

The CPAS must record any changes made to it or to devices and applications. Time-stamped audit trails show managers as well as inspectors:

- When changes were made
- Which changes were made
- Who made the changes, and why

The CPAS needs to make automation secure and track changes automatically. An integrated system-wide audit trail should let authorized users view and filter:

- User login, logout and change events
- Configuration changes
- Operator actions and inputs
- Alarm events
- System and device events
- Calibration alarms and events
- Batch alarms and events
- Signature events

All time-stamped audit trail events contain information about the object or file changed, the change itself and the action's owner. A CPAS should also log the reasons for the change and optional comments that the user has entered. Critical operations in a process or critical changes of machine settings might require significant authentication steps to ensure that the correct persons have the rights and authority to perform certain actions.

Manufacturers who decide to use additional security checks before users are allowed to download set points or issue commands rely on the HSI (human-systems interface). The user is required to provide username, password and new value and, if necessary, comments or a supervisor's signature before the change affects the process.

An electronically recorded audit trail (Figure 2–53) records all relevant changes during production, such as configuration changes or recalibration of an instrument. An audit trail helps to prevent personnel from making changes without being detected.

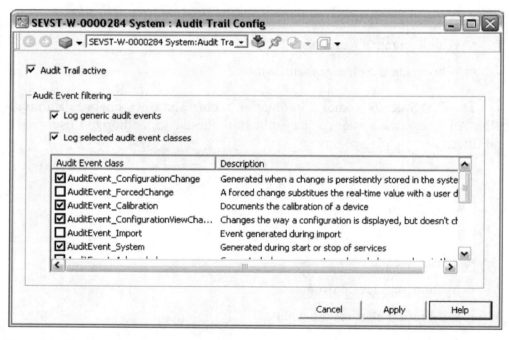

Figure 2–53   Audit Trail Configuration

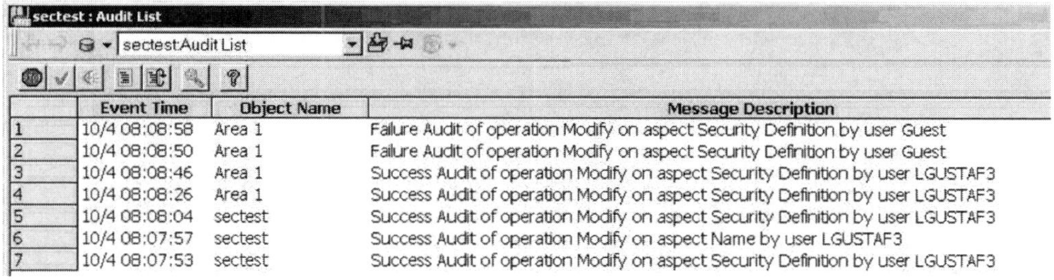

	Event Time	Object Name	Message Description
1	10/4 08:08:58	Area 1	Failure Audit of operation Modify on aspect Security Definition by user Guest
2	10/4 08:08:50	Area 1	Failure Audit of operation Modify on aspect Security Definition by user Guest
3	10/4 08:08:46	Area 1	Success Audit of operation Modify on aspect Security Definition by user LGUSTAF3
4	10/4 08:08:26	Area 1	Success Audit of operation Modify on aspect Security Definition by user LGUSTAF3
5	10/4 08:08:04	sectest	Success Audit of operation Modify on aspect Security Definition by user LGUSTAF3
6	10/4 08:07:57	sectest	Success Audit of operation Modify on aspect Name by user LGUSTAF3
7	10/4 08:07:53	sectest	Success Audit of operation Modify on aspect Security Definition by user LGUSTAF3

Figure 2–54 Example Audit List

Besides data integrity, audit lists (Figure 2–54) can be also used for other purposes. With the help of different filter configurations, lists supporting different tasks can be prepared. For example, an audit list can be a useful help to see what changes the last operator on shift made in the system.Another audit list can show all controllers that have been switched into manual mode, which is often an indication of suboptimal tuning of the controller.

Bibliography

Farb, D. and Gordon, B. *Agent GXP FDA Part 11 Guidebook*. University Of HealthCare, Los Angeles: 2005.

U.S. Food and Drug Administration *FDA 21 CFR Part 11* http://www.fda.gov/cder/gmp/index.htm.

CHAPTER 3

3.1 Engineering

Georg Gutermuth

Engineering is a craft that puts technical ideas into reality.

Introduction

In the context of this book, an engineering project could be building a new power plant (to meet the rising energy demands of a country), a metal rolling mill (to form metal slabs into sheet metal for cars), or a pharmaceuticals plant (for the production of a new medicine). The realization is finished when a gas fired power plant generates 500MW of electrical energy, a hot rolling mill produces 200 coils of sheet metal (14 tons each) per day, or a pharmaceutical plant produces industrial amounts of an active substance in the form of a powder that is harvested from large tanks of bacteria and used further in drug production. Examples of such engineering projects are shown in Figure 3–1.

Figure 3–1 Results of Large Engineering Projects: Power Plant (Europe), Mining Site (Venezuela), Oil Rig (North Sea)

3.1 Engineering

All plants have in common that, besides fulfilling their function highly efficiently, they must comply with all legal rules, be safe for the environment, and be a place where people can work safely. All this requires electric power, heat, telecommunication, information, climatization, waste disposal, transportation infrastructure and, of course, the necessary process ingredients.

In general, fulfilling any of those needs requires competence in understanding and capturing the requirements, planning and designing, realization of the plans in software and hardware, and finally testing and starting up (commissioning).

When one thinks of the many different people that are needed to build a private house, it is obvious that building an industrial plant requires many more people to be involved, each having special technical competencies to fulfill their tasks (trades). But at least at the same level, they need the ability to communicate, to understand the dependencies (interfaces) to other people's work. In addition, they must deal with large amounts of data, have a thorough and careful way of working to enable change management and versioning, understand cultural differences, and have the determination to cope with repetitive as well as challenging non-standard situations.

The tension and happiness of hundreds of people who have contributed to a project in the moment when a new plant starts production are indescribable.

This chapter focuses on a general overview of Engineering and discusses its non-technical aspects as well.

CPAS in the Context of Plant Engineering

Automation System Engineering is always embedded into a more complex series of work steps.

The decision to build a new plant is usually made by the company that wants to finally own it. This company is called the "owner/operator," and might decide on an external company to execute the complete building process on its behalf. Such a company is called an EPC (engineering, procurement and construction) contractor. The EPC contractor decides which parts (trades) of the work to execute himself and which will be executed by subcontractors. In many cases the subcontractors again have partners that they outsource parts of their work to. The automation system supplier is usually in the role of a subcontractor.

To get a good understanding of the various hardware and software engineering steps required for the automation system, it is important to look at its position within the overall system design. Even though different possible subdivisions of projects are possible, the eight shown in Figure 3–2 are commonly used.

In the first phase, called "front end engineering and design" (*FEED*), the owner/operator decides about technical details of the process (P&ID—piping and instrumentation diagram, safety), legal implications (location, contracts), and organizational matters (finances, timing, project partners, the bidding process).

All the following phases are usually highly parallel, and influence each other. They start with the spatial parameters during the *Site & Buildings* layout (earth movement, foundations, and streets) and the planning of the buildings (walls, ceilings, and metal structures).

In parallel to the building details, the (primary) *Process Equipment* that is needed for the main purpose of the plant is laid out. Examples from the different industries are turbines, generators, grinders, distillation columns, reactors, furnaces, steel rolls, boilers, fermenters, presses, and many more.

Once the details and location of the process equipment are decided upon, the passive devices (*Transportation & Storage*) can be planned to connect and support the process equipment. Examples are tanks, pipes, and conveyor belts.

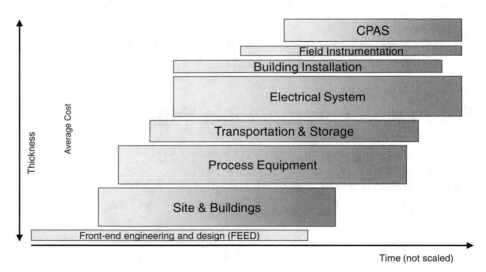

Figure 3–2   Schematic Time short of all Work Phases in a Green-field Plant Project. Indicated for Each Trade Are: Start Time (left x-position), Duration (x-extension) with Planning/Design (white) and Manufacturing/Configuration/Installation (grey). The y-thickness Indicates the Relative Cost Share, as Taken from Gutermuth (2007).

3.1 Engineering 159

In a later step the planning of the secondary equipment, needed for electrification, control, and support for humans, starts. Secondary equipment includes the *Electrical System*, comprising switchgear, transformers, uninterruptable power supplies, drives, electrical cabling, grounding, lightning protection, etc. *Building Installation* is an umbrella term for the supply and distribution of water, light, air, heating/cooling, networks for IT, sanitary facilities, communication (telephones, loud speakers) and security systems (cameras and intruder alarms).

Some time after those phases begin, the engineering for the automation system, that will measure and control the process parameters, starts. This phase comprises two parts: the *Field Instrumentation* (sensors for pressure, flow, level, temperature, mass, analyzers and actors such as pumps or valves) including field communication (such as fieldbusses) and the *CPAS* (control system, information management, batch control, logging, and advanced solutions such as MES (Manufacturing Execution Systems) and PPS (Production Planning Systems).

The classical term I&C has long been used to mean "(Field) **I**nstrumentation & **C**ontrol." In the context of this book, I&C is expanded to "(Field) **I**nstrumentation & **C**PAS."

The fact that the automation system is designed at quite a late stage results in the challenge that all the requirements and design changes (including mistakes) of the other phases have to be taken into account. Additionally, CPAS, in spite of being the newest technology, has rapidly developed into being *the* window to the site and the process. Naturally, the requirements for integration with Building Installation and Electrical expand the challenges, increasing interdependencies and other requirements and thus add to the complexity of the engineering of the CPAS. This explains why today I&C makes up 10% to 20% of the total project cost for a new plant (see Figure 3–2).

Factors Influencing I&C Engineering

The basic structure of a CPAS in all process plants is comparable. However, every plant is unique. Therefore, I&C projects, being "systems business", require more-or-less individual solutions to take care of specifics of each project. These include:

> **Green or brown field**—Only half of worldwide revenues in I&C projects are with completely new installations (called "green field" projects). Most of the projects are about modernization, retrofit, adaptation

or enlargements of existing plants (called "brown field"[1]). Brown field projects usually only apply to parts of a complete plant (e.g., HMI), while leaving most of it unchanged (e.g., field instrumentation). However, additional effort is necessary to update the documentation to reflect all changes in the plant over the past years of operation and modification.

Plant Size—The size of a plant can be measured by its spatial size (compare pipelines and oil rigs), by its construction cost (that integrates over several dimensions of complexity) or by the number of I/O (input/output) points. In first order, this number corresponds to the number of tags, process points or devices (either analog or digital). On a more detailed level, it is necessary to differentiate hard-wired I/O, Fieldbus I/O, and soft I/O from packaged units such as OPC servers. The median size of all I&C projects is about 1,000 I/O, with a range from <100 for small extension projects, to refineries being among the largest projects, with around 100,000 I/O in 2007. A trend towards larger and larger projects is ongoing. This chapter focuses on projects well above 1000 I/O, which require efficient bulk data handling (see Figure 3–15) and parallel multi-user (e.g., 10 users) engineering.

Scope of delivery—The EPC contractor can outsource anything from a small part of the control system to the complete I&C and electrical systems to one subcontractor. Naturally, a project with a larger scope bears more optimization potential and a different workflow than a very limited one with rigid boundary conditions that might, however, need less time to execute. Even so, the tasks that must be performed stay the same, regardless of who performs them.

Plant location—Another influencing factor is the location of the plant. One can imagine that the testing of an I&C system for an off-shore oil rig will mainly be done on land; hydroelectric power plants in the middle of nowhere will use many pre-assembled parts, and countries can have extremely varying regulatory requirements on environmental certificates and other local laws to conform to.

---

1. Not to be confused with "brown field" projects that are undertaken to remediate soil and groundwater pollution.

Company requirements—Most owner/operators, as well as the I&C subcontractor, will have requirements according to their business strategy and preferences (e.g., choice of hardware, protocols, or tools). Large automation companies might use a globally distributed team of specialists, whereas smaller companies might have a team concentrated at one location.

Domain—A factor that can make a huge difference is the domain. In the remainder of this chapter, we will consider only those requirements that are common to all process industries and utilities projects. However it is interesting to mention some of the specifics:

o **Oil & Gas** installations deal with explosive material and thus have strong requirements for safety (see Chapter 2.8). The fraction of the (mainly digital) safety I/O for fire and gas detection can be as high as 50% of all I/O. That is one reason why refineries are among the largest plants for I/O count. Off-shore installations (oil rigs) are very little standardized due to the variety in oil composition and sea(bed) conditions. All, however, have very limited space, resulting in a high packing density of the hardware. Pipelines, on the other hand, can stretch thousands of miles with the need for remote monitoring.

o **Pharmaceuticals** as well as the **Food & Beverage** industries face high demands on information traceability. Requirements of the Food and Drug Administration (FDA) can result in 30% of personnel work time being spent on documentation! The diversity of the products results in very individual and sometimes flexible plants. As the final product is usually some kind of batch (bottle, pill box, etc.) the processes are semi-continuous and thus the control code has a high ratio of control sequences for batch control.

o **Chemical plants** have a high number of sensors and analyzers for product quality and emission monitoring and thus 50% of the data transmitted is analog process values. The diversity of the products (and the plants) can be very high. As the process is completely sealed, the "buildings" often degenerate to open metal structures with pipes attached to them. Safety requirements are increasing recently. The "safe" state of a chemical plant might not be to switch it off, as in other plants, but, e.g., to actively cool a vessel to stop the chemical reaction in it.

- **Sheet processes (for paper or metal)** have high-quality requirements combined with high speeds, resulting in the highest demands on fast (millisecond) on-line measurement (e.g., thickness), control (e.g., roll pressure) and communication. Advanced process simulation and virtual commissioning require specific know-how of the process, even though it is highly standardized.
- **Mining** is about prospecting for, obtaining, crushing, and transporting ore or stone. As the spatial dimensions can be very large, the need of transportation-actuators (motor drives), communication (wireless) and localization (GPS) is evident. The solution itself can be highly standardized.
- **Power plants** naturally have an above average amount of electrical equipment. Upcoming are new requirements on emission monitoring, demand driven production scheduling, as well as safety (e.g., for boilers or turbines). The diversity among, e.g., coal-fired power plants can be very small and thus a high degree of standardization is possible.
- **Utilities** such as (waste-) water transportation or treatment (e.g., desalination), can have a large spatial distribution. This results in a large number of actuators (e.g., pumps) and a demand for remote monitoring and operation, including some unmanned substations or sewage plants.

Despite all the differences, the basic tasks in all mentioned domains are still so comparable that general standards for communication, visualization and engineering have become widely accepted, and automation system vendors are converging towards supplying the complete range of industries with one set of hardware and tools.

I&C Engineering Tasks

Many sources describe engineering phases that are similar to the ones in Figure 3–3. For a basic understanding, this view is very good and helpful. Usually, phase 1 is executed by the EPC contractor and the result of phases 1 to 3 is a set of documents that are used (and completed) during the following phases. Phases 2 to 4 are completed at the site of the automation system supplier (or its sub-suppliers), whereas 6 and 7 have to take place at the (final) plant site.

3.1 Engineering

1. Basic Determination	2. Preliminary Engineering	3. Basic Engineering	4. Detail Engineering	5. Construction	6. Commissioning	7. Project Completion

Figure 3–3 Seven Phases of Project Engineering According to NAMUR Worksheet 35 (NAMUR NA 35, 2003)

Phase 5 mainly takes place at the plant site with pre-assembly taking place at other sites.

However, to get to a more detailed level, the ideal case of a linear order of actions has to be abandoned. Figure 3–4 gives an overview on 17 tasks that are executed in a highly parallel fashion during I&C (green field) project execution. They are mapped to five phases that are slightly adjusted, in comparison to Figure 3–3.

Figure 3–4 structures projects according to "tasks" and thus the education and skills of the person performing the task. Typical large projects require sales, a project manager, a lead engineer, procurement, IT specialists, Fieldbus specialists, control system application engineers, operator graphics engineers, technicians, a commissioning engineer and others. Not all people are part of the team all the time. People might join or leave as required (with a peak during Detail Engineering), some are never directly part of the project team (e.g., procurement), while others might switch roles during the project (e.g., from application to commissioning engineer).

For a better understanding we will explain each of the 17 tasks in more detail. Whitt (Whitt, 2004) gives a comprehensive and more detailed description of not only what to do, but how.

Offer Preparation. As a basis for I&C projects, the EPC contractor provides requirements from the trades: FEED (PFD[2], P&ID[3]-diagram, study results (e.g., HazOP = Hazard and Operability Study), instrument list and number of I/O, "Site & Buildings" (first 3D layout) and "Electrical System" (motor and component list, single line diagram and requirements for integration of third- party systems[4]). This information, together with other requirements, is called the "user requirement specification" (URS)[5] and serves a basis for an initial system

---

2. PFD = Process Flow Diagram
3. P&ID = Piping and Instrumentation Diagram
4. Examples of third-party systems (so called "packaged units") are safety systems, turbines, gas analyzers, fire & gas systems or diesel generators.
5. Sometimes called "performance specification"

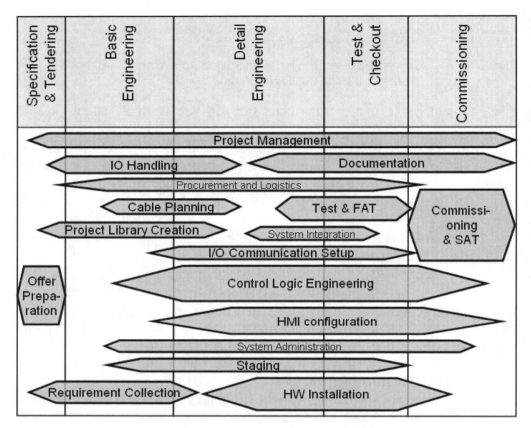

Figure 3–4 Schematic Chart of an I&C Project Plan. (Note: The x-position of each task indicates its duration and timing with respect to the five project phases. The size of each task area indicates the effort needed. The y-position is random.)

partitioning (number of controllers), network topology, bill of material, effort estimation, project plan and finally the offer price.

Requirements Collection. Once the sub-supplier has won the bid they must come to a mutual understanding with the owner/operator about the detailed requirements and an agreement about certain design decisions. This is usually done in a series of meetings, resulting in detailed supplier documentation called the "functional design specification" (FDS), which is usually an addition to the contract. Examples of decisions made are control libraries to be used (including graphical appearance), safety requirements, amount and layout of end-documentation, test specifications, control-room layout, operator display

3.1 Engineering 165

layout (composition, content and style guide), control strategies, and detailed control description (everything from verbal to formalized).

Project Management. Large automation projects can be extremely challenging from a management point of view. Therefore, professional project management is essential. It contains project planning (resources, timing, dependencies), project communication (status tracking, internal and external reporting), project controlling (quality management, deviations from plan, risk management, claim and change management), and contract handling (suppliers, subcontractors, customer). Modern engineering environments support parts of those tasks, e.g., change management or status reporting (Liefeldt, 2005).

Project Library Creation. An important factor for efficiency in engineering, both hardware and software, is the reuse of standard "objects" (called types, typicals or templates). They are collected, modified or created for a project-specific library. In the case of control software, the functionality (PID, motor function block, switchgear-connector, burner control) must comply with the customer requirements and the graphics (faceplate, icons) with the accompanying guidelines. Already existing "standard" libraries need to be adjusted and tested. Other typical examples are control-rack layout, documentation layout, or cabling typicals. All elements must often be approved by the owner/operator.

I/O Handling. Basing on the P&ID or an initial tag list, an I/O list[6] is compiled. Starting from initial information (tag name, description, physical unit and value range) the list is constantly updated and filled with more information: alarm limits, HW (hardware) realization, SW (software) realization, and cycle time as well as assignments to controller task, control typical, I/O-channel number or IP address on the communication bus. Each additional piece of information involves many design decisions and good change management. With the large amount of data, list-based representation and manipulation have proven to be the most effective. An example is shown in Figure 3–5. When it comes to single decisions, graphic-oriented views are preferred.

Cable Planning. A physical connection between the I/O channel of a controller and the field device is rarely one single cable. The information, in the classical case (4-20mA), is routed through junction boxes, marshalling panels, cable trays and conduits or might include fieldbus communication as connection to remote I/O. This so-called "*inter*-cabinet wiring" may pass different

---

6. In projects, the terms I/O-list and Tag-list are not always used consistently. Additionally, other names may be used, such as signal list or instrument index.

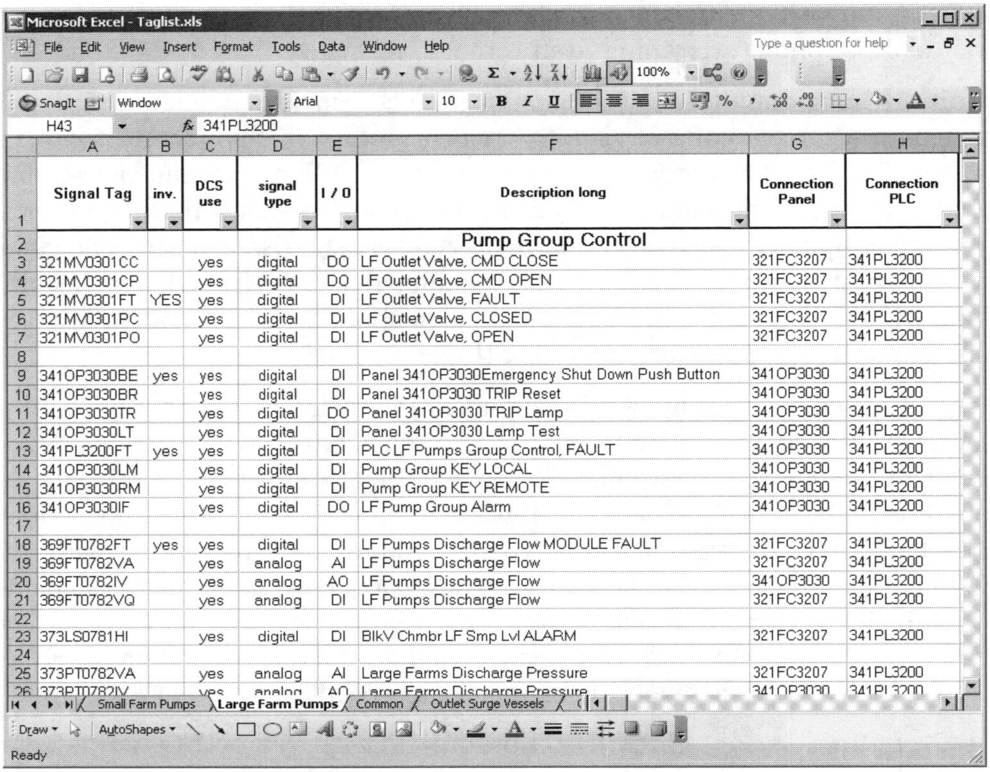

	Signal Tag	inv.	DCS use	signal type	I / O	Description long	Connection Panel	Connection PLC
2						Pump Group Control		
3	321MV0301CC		yes	digital	DO	LF Outlet Valve, CMD CLOSE	321FC3207	341PL3200
4	321MV0301CP		yes	digital	DO	LF Outlet Valve, CMD OPEN	321FC3207	341PL3200
5	321MV0301FT	YES	yes	digital	DI	LF Outlet Valve, FAULT	321FC3207	341PL3200
6	321MV0301PC		yes	digital	DI	LF Outlet Valve, CLOSED	321FC3207	341PL3200
7	321MV0301PO		yes	digital	DI	LF Outlet Valve, OPEN	321FC3207	341PL3200
8								
9	341OP3030BE	yes	yes	digital	DI	Panel 341OP3030Emergency Shut Down Push Button	341OP3030	341PL3200
10	341OP3030BR		yes	digital	DI	Panel 341OP3030 TRIP Reset	341OP3030	341PL3200
11	341OP3030TR		yes	digital	DO	Panel 341OP3030 TRIP Lamp	341OP3030	341PL3200
12	341OP3030LT		yes	digital	DI	Panel 341OP3030 Lamp Test	341OP3030	341PL3200
13	341PL3200FT	yes	yes	digital	DI	PLC LF Pumps Group Control, FAULT	341OP3030	341PL3200
14	341OP3030LM		yes	digital	DI	Pump Group KEY LOCAL	341OP3030	341PL3200
15	341OP3030RM		yes	digital	DI	Pump Group KEY REMOTE	341OP3030	341PL3200
16	341OP3030IF		yes	digital	DO	LF Pump Group Alarm	341OP3030	341PL3200
17								
18	369FT0782FT	yes	yes	digital	DI	LF Pumps Discharge Flow MODULE FAULT	321FC3207	341PL3200
19	369FT0782VA		yes	analog	AI	LF Pumps Discharge Flow	321FC3207	341PL3200
20	369FT0782IV		yes	analog	AO	LF Pumps Discharge Flow	341OP3030	341PL3200
21	369FT0782VQ		yes	analog	DI	LF Pumps Discharge Flow	321FC3207	341PL3200
22								
23	373LS0781HI		yes	digital	DI	BlkV Chmbr LF Smp Lvl ALARM	321FC3207	341PL3200
24								
25	373PT0782VA		yes	analog	AI	Large Farms Discharge Pressure	321FC3207	341PL3200
26	373PT0782IV		yes	analog	AO	Large Farms Discharge Pressure	341OP3030	341PL3200

Figure 3–5 List-based I/O Handling in Excel (small part of a tag list of an example project with 2000 I/O)

safety or hazardous zones and be part of various bundles of cables. Things to consider are shielding, cable lengths, resistance/damping, color codes, materials and cost. Fieldbus communication requires additional considerations. The cabling within the cabinets is considered as part of the Staging (see staging section).

Procurement & Logistics. Basing on the requirements from the URS/FDS, the bill(s) of material can be compiled. Choices concerning the specific types have to be made and suppliers need to be found, not to forget the ordering, billing, transportation, logistics, storage, and possible reclamation processes involved. This is true for very different HW types: field devices, control and I/O equipment, communication hardware, PCs, control room equipment (monitors, displays, chairs, tables, etc.), power supplies, housing, cabinets and bulk commodities (e.g., mounting material or cables). Parts of the control hardware are shown in

3.1 Engineering

Figure 3–6 Example of Control System Hardware (monitors, keyboard, HMI-panels, controller, I/O-boards, communication interfaces, CD-ROM and mouse)

Figure 3–6. However, some parts of the hardware are first shipped to engineering companies for assembly, installation, programming and/or testing. Other parts are shipped to the I&C supplier or directly delivered to the plant site.

System Administration. Once the ordered IT equipment (server, clients, monitors, printers, software and licenses) arrives on the engineering site, the configuration of the PCs may require installing the operating system, standard (office) software, security software and the control engineering software—they even come preinstalled in some cases. Furthermore, modern multi-user systems require setting up a computer network (servers, clients and control equipment), backup mechanisms, and complete user and access management. It can be an advantage to configure views, access rights and available functionality within the engineering environment. During the project's duration, system re-configuration or software upgrades might be required.

Staging. Once all design and partitioning decisions have been made, the layout of the control cubicles, marshalling racks and possible remote I/O cabinets can be planned. That involves the controllers, I/O boards, power supply, cabling, terminal units, and mounting as well housing equipment. After delivery of the needed hardware, the setup (= staging) of the cabinets starts, based on (most of the time, graphical) design documents. The interior of a control cubicle can be seen in Figure 3–7. Example numbers are 33 control cubicles and 15 marshaling racks for a 25,000 I/O coal fired power plant. Before shipment to the plant site, a cabinet test and verification are performed.

I/O Communication Setup. Depending on which communication protocol is used, this task requires more or less work. For 4-20mA signals (which, including

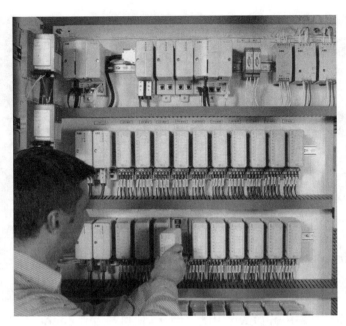

Figure 3–7 Interior of a Control Cabinet After Staging is Finished (Note: controller, power supply, and local I/O units are mounted and cabled)

HART protocol, are >50% of all signals today), the cable routing and I/O allocation are the only information needed (→ "I/O Handling" further up). Some configuration of the I/O channel units might be needed to adjust value ranges or sampling times. Modern fieldbus communication such as PROFIBUS or Ethernet-based protocols (FOUNDATION Fieldbus, PROFINET) offer more functionality and flexibility but, on the other hand, require more configuration: planning and setting up the communication network, assigning IP addresses, choosing device operation modes, loading device functionalities and interfaces (e.g., DTM—device type manager). Further, more specific communication to "packaged units" needs to be set up (e.g., via OPC or PROFIBUS Link). A tool example is shown in Figure 3–8.

HMI Configuration. The operator workplace usually consists of many (dozens to hundreds) process displays with different hierarchical levels and granularity (from overview and schematics to detail displays) and access to additional functionality such as Alarm & Event, information management, trending, logging, asset management and many more. As it is the part of the automation system that is most visible to the owner/operator, many requirements are usually to be found in the URS about the display hierarchy, layout, use-cases and color codes.

3.1 Engineering 169

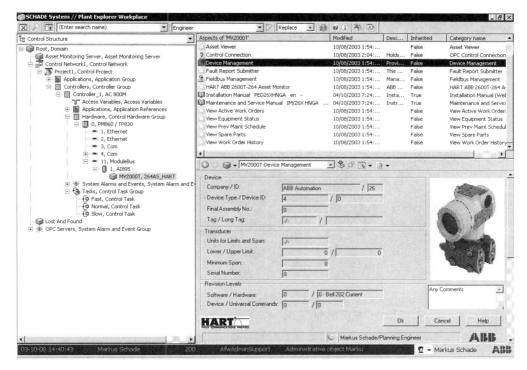

Figure 3–8   Example of a Tool for Device Configuration

The pure generation of the single graphical elements can be automated. The placement and sizing within the process graphics, however, are done manually in the light of ergonomics and usability for the operator. Furthermore, navigation links and the setup of all mentioned functionality have to be provided. A final operator screen can be seen in Figure 3–9.

Control Logic Engineering. The task of implementing the control code software according to the specifications (URS, FDS, I/O list) is the core and thus largest task of I&C projects. It is a current trend to keep the implementation of the functionality neutral from a specific hardware realization as long as possible (this is called function oriented top-down engineering).

Code generation can be roughly separated into three parts that are executed in one tool (e.g., as in Figure 3–10):

- Create tag and I/O objects in the control environment. This has mainly become a matter of auto-generation of typicals (copying) or types (instantiating).

Figure 3–9 Operator Workplace (on the displays) with Large Overview Graphic and Smaller Detail Graphic Displays

- Parametrizing the (large) objects and connecting them with so-called "interlock logic" or "glue logic." This code can be analog (e.g., process control) or digital (e.g., safety logic).

- Implement the "steering logic" or "sequence control." Examples are batch production, plant startup or emergency shutdown sequences.

Depending on I/O count of the project, one person might only implement a sub-part of the plant, called a "functional unit," consisting of digital, analog and/or sequence logic.

System Integration. Parallel engineering of different tasks requires largely uncoupled detail engineering (Control Logic, HMI, Field Communication and Control HW/Staging), with a later integration of the part-results. This task is a sum of separate integration steps:

- Connecting *Communication* and *Control HW* requires setting up a physical connection (cable) and some parameter settings.

- Integrating *Control Logic* and *HMI* requires connecting the SW signals with the corresponding variables in the HMI.

3.1 Engineering

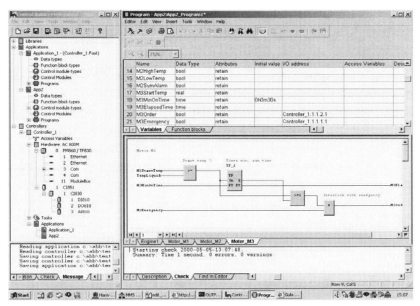

Figure 3–10 Control Application Environment with Project Tree (left), Variable List (top-right), FBD-editor (middle-right), and Message Windows (bottom-right)

- Connecting *Control Logic* with *Control HW* requires setting connection parameters and assigning the applications to tasks within a specific controller.

- Finally, the I/O allocation between *Control Logic* and *Communication* requires matching the physical device to I/O channel allocation with the pure software allocation of SW signals to SW channels.

When all four separate tasks have worked on a consistent version of the I/O List, this task could become completely automatable. However, in reality constant changes in the data usually prevent this.

Test and FAT. All the above tasks already include some kind of tests. Once Detail Engineering and System Integration are finished, functionality and interplay of Control Logic, HMI, Communication and Control HW are tested more formally, before shipping the solution to the customer's site. The so-called "checkout" is started with a System Integration Test (SIT) for basic functionality, such as connectivity, redundancy, time synchronization, user settings, memory, and CPU load. The second step, called Functional Test, ensures the correct behavior of the control logic, interlocks, sequence control, graphics, process alarms, and logging and trends. The last step, called the Factory Acceptance

Test (FAT) is a formal verification according to a checklist, as specified by the EPC contractor. After successful completion, all hardware can be shipped to the plant site.

Installation at the Plant Site. All hardware will finally be installed at the plant site according to layout drawings and cable plans: Field devices (sensors and actuators) are connected to the process equipment (mounting and hook up, see Figure 3–11); the field cabling from the devices to junction boxes and all the way to the control cubicles is pulled and connected; the control cubicles and marshaling racks are erected and wired for (external) electrical and control connection, and the control room with all its equipment (furniture, screens, servers, PC clients, network equipment) is set up.

Commissioning and SAT. In the final project step of putting the plant into operation, many of the phases in Figure 3–2 are involved: process equipment, electrical system, building installation, field instrumentation and CPAS. The CPAS, however, as "the window to plant and process," plays a dominant role during commissioning. First, the control system itself must be commissioned (server redundancy, communication, access rights). The "field hardware test" of the communication system can be performed in parallel, and the connection of all "packaged units" to the control system must be completed.

The next test, called "loop-check," verifies all loops—namely the complete path from a field device to the HMI (including the device, cabling and communication network, I/O and controller hardware, control system, and operator station). The test is carried out by forcing a signal at one end of the loop (e.g., increasing the heat at the temperature transmitter or pushing the "close" button on the valve faceplate) and confirming the reaction at the other end.

After completion, the "cold commissioning" can start. It implies running the plant with many test sequences and nearly full functionality, but with reduced risk (e.g., water instead of chemicals). This is a good point in time for parallel control loop tuning and operator training.

Once all tests have been passed successfully, the final "hot commissioning" brings the plant into operation and production. Handover to the owner/operator is finalized with the site acceptance test (SAT).

Documentation. The amount of final (or "as built") documentation depends on the requirements of the owner/operator and legal regulations. The documentation can include the URS, control philosophy description, control diagrams, wiring plans, description of the HMI, maintenance information, test and approval records, communication network topology, bills of material, and all hardware and software

3.1 Engineering 173

manuals. The format of the documentation used to be literally meters of folders full of paper printouts. Today there is a trend towards electronic formats like .pdf and more recently, (additional) object-oriented plant models are asked for.

In summary it can be said that each phase of Figure 3–4 can be defined by a certain work result to be achieved. This is illustrated in Figure 3–12.

The final result of all engineering steps is a plant that can be monitored and controlled from a central operator room with one or very few people, as shown in Figure 3–13.

Figure 3–11 Installation of Field Instrumentation in an O&G Installation (Mexico)

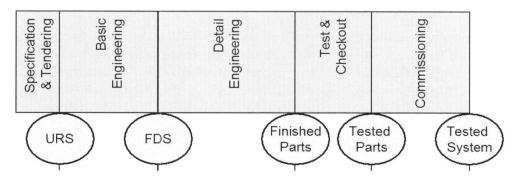

Figure 3–12 Work Result at the End of Each Engineering Phase

Figure 3–13 Operator Room in a Waste-to-Energy Plant in Sweden

Bulk Data Handling

"Proper management of large amounts of information has always been a major part of the I&C profession." (Whitt, 2004) As already mentioned, projects can have several 10-thousands of I/O. So it is evident that a graphical representation and manipulation of single objects quickly reaches its limits. Therefore, effective bulk data handling methods are needed. The tasks can roughly be divided into three use-case groups:

I/O Handling Support. The goal and tasks of this category have already been described in the previous section. Let us consider the following example: the question "Which control objects implement which tags?" has already been decided. For each control object, the cycle time is already decided. In the next step, each control object will be assigned to a certain task on a certain controller. This could be realized by the following steps:

- Filtering for all control objects that belong to the same functional unit and thus should be executed on one controller. The common attribute to look for could be a sequence in the Tag name or control object identifier.[7]

---

7. For example, the KKS (Kraftwerk-Kennzeichen-System), widely used in Europe.

3.1 Engineering 175

- Once the selection has been made, choosing a controller name or ID from a list once should be sufficient to assign the whole selection to this controller.

- Now we can filter again, this time for all control objects (a) on the chosen controller and (b) with a cycle time within a certain range.

- Assuming that we want the first half of the selection to be in the same task, we could type the task-name once and auto-fill for the selection we have made.

The important features are thus search/query, cross-referencing, filtering/(multi-)selection, auto-completion, extensibility, and proposing possible entries. Popular tools that offer those functions are based on spreadsheets (e.g., Microsoft Excel) or databases (e.g., Microsoft Access or SmartPlant Instrumentation).

Copy and Cloning. In all plants there are functional units and other things that are repeating (an example is given in Figure 3–14), so why engineer them twice?

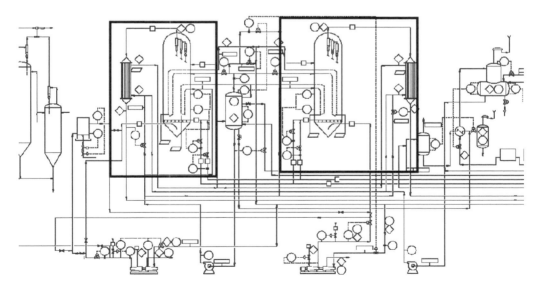

Figure 3–14   Process Flow Diagram (PFD) with Two Very Similar Units (marked with black rectangles)

However, copying them involves more than just copying text or tables:

- Even with identical solutions, at least the names must be unique. The tag, faceplates etc. of the second unit might have a number "2" instead of "1" in their identifier. The controllers, on the other hand, might just be numbered continuously: when unit one uses controllers 1 to 8, the clone would need controllers 9 to 16. Thus efficient and intelligent *renaming* is needed.

- The *external interfaces* have to be found and either disconnected or intelligently re-connected. For the field devices there might be a second, identical set of instances (with other names). The "emergency shutdown signal," however, might be unique for the whole plant. The detail operator displays will be cloned, but for the master overview the two identical units might be on one page.

- Aspects of the copied instances might need to be handled differently. The creation date/time would need to be adjusted. Maintenance documentation for identical hardware could just be referenced. References to object *types* can be copied. Settings for redundant equipment would need to be changed from "master" to "secondary master."

A smart copying function is a basic functionality of software engineering tools. However, to be able to clone parts of the plant efficiently the master copy must be designed intelligently (naming schemes etc.) and the rules to be applied should be automatable by the software as far as possible.

Special Tasks. Besides the two mentioned standard bulk operations for "I/O Handling" and "Copy and Cloning", there can be literally hundreds of additional special use-cases. To mention only five examples:

- During FAT it is realized that one controller is overloaded. Therefore, an additional controller will be introduced and parts of I/O and application have to be reassigned.

- A slightly different pressure transmitter is being delivered than planned. Therefore, all instances have to be found and corrected accordingly.

3.1 Engineering 177

- During testing many values have manually been set (forced) or simulated. For commissioning, all forced and simulated signals must be found and set again to automatic mode.

- A list of all PROFIBUS device locations might be needed for verification by the appropriate specialist.

- After receiving a revision of the I/O list, a difference report on the implemented installation is needed.

Some of those special tasks can efficiently be handled in the same tool as the I/O handling. Others will need to be implemented during the run of the project. Good programmers' interfaces for data searching/filtering, export, manipulation and import are very helpful. See Figure 3–15 for a Microsoft Excel-based tool.

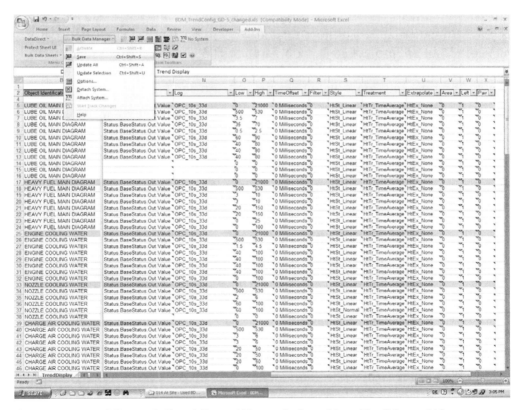

Figure 3–15 Microsoft Excel-Based Example for a Very Flexible and User-adjustable Bulk Handling Tool for "Special Tasks

Challenges and Trends in I&C Engineering

The CPAS (and with it, I&C engineering in general) has undergone major changes in the past, and current trends will continue to change functionality, handling and appearance. We will briefly discuss six trend-clusters, listed in Figure 3–16 together with the reasons of the stakeholders for promoting them. For details, please refer to Früh and Maier (Früh, 2004).

Reduce Engineering Cost. Engineering has developed from a necessity to a differentiating factor. The cost of engineering (all project-related work) has constantly increased to >50% of total I&C project costs, compared to HW and SW. Interestingly, the reason for this is technological progress. New technologies (network IT, fieldbusses), advanced functionality (asset or alarm management), and increasing requirements (flexibility, documentation and safety) all need to be engineered. The cost for the hardware on the other hand was drastically reduced through the usage of standard components. This is why the cost of Engineering (all project related work) constantly increased to around 50% of total I&C project costs, compared to HW and SW. Additionally shorter cycles for I&C products, plant updates, and even the engineering workforce demand engineering tools that are easier to learn and use.

Better Tool and Data Integration. Integration of all tools and data in three dimensions is a strong trend. *Vertical Integration* is naturally realized from plant floor to HMI, but is currently being expanded to the management level for production optimization and management and decision support. Examples of functionality to be integrated online are AM (Asset Management), MES (manufacturing execution system) and ERP (enterprise resource planning). *Integration between trades* (see Figure 3–2), simplifies the task of the EPC contractor and opens doors for further plant optimization (e.g., production output vs. energy consumption). Finally, *Horizontal Integration* between project phases from the FEED phase (CAE; computer-aided engineering tools) along the engineering life cycle in Figure 3–3 (Engineering tools) to the operations phase (HMI) saves manual retyping and supports automatic change management. Promising approaches are the standardization of syntax and semantics like CAEX (Ebel et al., 2008) or NE100 (NAMUR, 2007).

Increase Standardization.

- **Methods**—Standardized data models and interfaces are key to tool integration. The standardization of engineering methods supports

3.1 Engineering 179

Trend of... \ Stakeholder	Owner/operator	EPC-contractor	Automation Supplier	Research
Reduce engineering cost to increase profit by...	reduced total project cost	higher margin in projects	stay competitive with low offer price	tool usability, automation of automation, global-sourcing
Better data integration to have single data entry ...	along lifecycle (horizontal) & from field to management (vertical)	across trades	along the engineering tasks (horizontal)	integration methods, onthologies, standards
Higher degree of standardization to...	be independent of supplier	make integration easier	reduce engineering cost (vs. keeping differentiators!)	SW methods in automation, reuse
Introduction of new technologies to...	improve operations & reduce maintenance	fulfill customer requirements	reduce engineering cost and fulfill customer requirements	agent-based automation, wireless, data-mining, optimization
Increased flexibility of the CPAS to...	enable flexible production & easy extendibility	-	offer differentiating features and have additional revenues	change management
Have a digital image of the plant to...	enable as-is documentation, staff training & production optimization	offer optimized solutions with a better quality	reduce engineering cost while optimizing result & quality	modeling & simulation, optimization

Figure 3–16 Reasons (white) for the Three Stakeholders (grey, top) to Promote Certain Trends/Goals (grey, left) in Automation Systems and the Corresponding Academic Research Fields (light grey)

usability of the tools and thus reduces engineering cost. Every engineering environment today offers implementation languages, similar to IEC 61131-3 (see Chapter 3-2). Formal descriptions of control code or tests, e.g., with Petri nets or UML (Unified Modeling Language), have not yet been adopted fully by industry.

- **Software**—Software for HMI and engineering largely converged to standards like Windows look-and-feel and Microsoft operating systems. Standard software for documentation (Word, pdf), list handling (Excel) or drawings (Visio, AutoCAD) are as widespread as OPC, XML or Web browser technology for communication and data exchange. As predicted in the 1990s (Song, 1996), features known from software development environments (e.g., Microsoft Visual Studio), are on the rise for control software development: object orientation for reuse, versioning concepts for the code, graphical engineering complements list-based methods, locking (check-in/check-out) for multi-user engineering, user group customization and rights management, syntax highlighting, Microsoft IntelliSense, and easy search and navigation.

- **Hardware**—Proprietary hardware components are being replaced by COTS (commercial off-the-shelf) for visualization (PC, monitor), calculation (PC or standard chip sets with firmware) and communication (Ethernet based).

New Technologies.

- **Software**—Today's CPASs extend the classical functionality by Asset Management, extended equipment information, optimization capabilities, enhanced Alarm&Event handling, model predictive control strategies, and interfaces to many other systems. This trend is ongoing, but the benefits of increased possibilities come with the downside of additional engineering effort.
- **Hardware**—The same is often true for new *Hardware*. New safety integrated systems or fieldbus protocols do usually not bring effort reduction in the early engineering phases.
- **Distribution**—A special trend is *Distribution*. Having started with a few central processing units (CPU), redundant and distributed control systems (DCS) soon emerged. Now with FOUNDATION Fieldbus, PROFINET, and IEC 61499 (Hall, 2007), the functionality can be distributed even further. Academia is currently researching completely agent-based control systems.

Increase Flexibility. Software product life cycles are decreasing and this requires more flexible automation solutions that can be changed and extended easily. And customization of functionality and appearance are increasingly requested: Alarm&Event, user settings, color codes or NLS (Native Language Support) might all be owner/operator specific.

Virtualization. Simulation of (parts of) the hardware offers many advantages: virtual controllers (soft controllers) enable testing even before the hardware is present and thus reduces calendar time. Commissioning of virtual Fieldbus networks in brown field projects reduces plant downtime. Complete plant simulation enables optimized decisions during the design phase. However, building the "digital plant" today still requires large amounts of additional work.

Conclusion

Engineering, both hardware and software, is a complex task due to large amounts of data, many technical and human interfaces, and constant changes. Whitt (Whitt, 2004) describes it as sometimes "more of an art than a science." Mastery of efficient engineering is a challenge, but promises competitive advantage by total project cost reduction and quality increase. Furthermore, engineering is often the key enabler for many new technologies and future requirements.

References

Ebel, M., Drath, R, and Sauer, O. "Automatische Projektierung eines Produktionsleitsytems der Fertigungstechnik mit Hilfe des Datenaustauschformats CAEX" In: atp—Automatisierungstechnische Praxis 50 (2008), Heft 5, Seiten 84–92. München, Oldenburg Industrieverlag.

Früh, K.F. and Maier, U. "Handbuch der Prozessautomatisierung" 3. Edition 2004, München, Oldenbourg Industrieverlag.

Gutermuth, G. and Hausmanns, Ch. "Kostenstruktur und Untergliederung von Automatisierungsprojekten" In: atp—Automatisierungstechnische Praxis 49 (2007), Heft 11, Seiten 40–47. München, Oldenburg Industrieverlag.

Hall, K.H., Staron, R.J. and Zoitl, A. "Challenges to Industry Adoption of the IEC 61499 Standard on Event-based Function Blocks" downloaded from IEEE Xplore on 25.3.2009, 1-4244-0865-2/07.

Laubner, R. and Göhner, P. "Prozessautomatisierung 2" New York, Berlin, Heidelberg: Springer-Verlag, 1999, ISBN: 3-540-65319-8.

Liefeldt, A., Gutermuth, G., Beer, P., Basenach, S., and Alznauer, R. "Effizientes Engineering—Begleitende Fortschrittskontrolle großer Projekte der Automatisierungstechnik." atp—Automatisierungstechnische Praxis 47 (2005), Heft 7, Seiten 60–64.

NAMUR NA 35 *"Abwicklung von PLT-Projekten / Handling PCT Projects"* NAMUR Worksheet NA35, version of 24.3.2003. www.namur.de

NAMUR *NE 100 "Use of Lists of Properties in Process Control Engineering Workflows."* Namur Recommendation NE100, version 3.1. of 2.11.2007. www.namur.de

Song, F.J. "Next generation of structural engineering automation systems." In Computing in Civil Engineering, (1996), (New York), pp. 494–500.

Whitt, M.D. "Successful Instrumentation and Control Systems Design" Research Triangle Park: ISA, 2004. ISBN: 978-1-5617-992-1.

3.2 Control Logic Programming

Georg Gutermuth

Introduction

The core activity of an automation project is the creation of the control functionality, as specified in the URS, FDS and other documents. For control code implementation, several alternatives, called *(programming) languages,* exist.

When programmable process control systems evolved in the 1960s-1970s, each hardware supplier developed their own language for implementing the functionality. Some programmable logic controller (PLC) vendors chose a programming style similar to relay logic, which was known by electricians; others used textual languages similar to early computer programming (Assembler or Basic). For discrete processes, the "control flow" was represented in a more graphical fashion. This resulted in a variety of programming languages.

However, for plant owner/operators it is beneficial to have the languages compatible and reduced in numbers. The advantages are reduced training effort for maintenance personnel, easier changes of suppliers and hardware, re-use of solutions across platforms, and simplification of migration between systems. For those reasons, in 1993 the IEC (International Electrotechnical Commission) published the 61131-3 standard, including five programming languages.

IEC 61131-3

IEC 61131 is an international standard about "programmable controllers" in a broad sense (definitions, requirements, tests, hardware, software and communication) consisting of eight parts.

Part 3, called "programming languages," defines five programming languages with syntactic and semantic rules, sets of programming elements, applicable tests, and means by which the basics may be expanded upon by manufacturers. For further reading, we recommend John and Tiegelkamp (John, 2001) or the standard itself (IEC 61131-3).

The five standardized control system programming languages have a textual, graphical or mixed appearance, as shown in Figure 3–17.

All five languages have in common the following five elements:

1. **Data Types** specify the (memory) size and meaning of variable contents. The main ones can be separated into four different categories. (The numbers in brackets give the memory consumption in Bits.)

    o *BIT Data Types*: BOOL (1), BYTE/CHAR (8), WORD (16), DWORD (32) and LWORD (64)

    o *Arithmetic Types*: SINT (8), INT (16), DINT (32), LINT (64), REAL (32) and LREAL (64). The Integer types can be unsigned as well e.g., UINT (8)

    o *Time Types*: DATE (16), TIME (32) and TIME_OF_DAY (32)

    o *Multi-element (alternatively composed or derived) data types* require the memory consumption to be specified during programming: ARRAY, STRUCT and STRING

2. **Variables** that need to be declared (data type, name or address), scoped (local, global, input, output, restart behavior) and initialized (default initial values exist). Example-declarations are shown in Table 3–1.

3. **Program organization units** for structuring the code:

    o *Functions* calculate and return exactly one result value, based on one or more input values. They are called within *function blocks* and have no internal memory. Functions are usually provided by the CPAS manufacturer as the smallest (functional) building blocks.

    o *Function blocks* can calculate and return more than one value and can store internal states even after the function block is being exited (memory function). A second call with the same input parameters may thus lead to a different result. Function blocks are instantiated within *programs* or other *function blocks* and are often user defined.

    o *Program* is the top level logical assembly of all elements and constructs necessary for the intended signal processing. They are called within *resources*.

3.2   Control Logic Programming   185

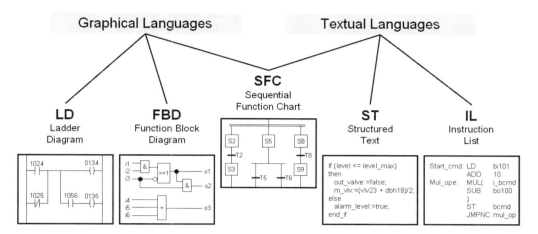

Figure 3–17   Classification of the Five IEC 61131-3 Programming Languages

Table 3–1   Example of Textual Declaration of Three Local Variables

Example	Explanation
VAR	
AT %MW48 : INT;	Allocates a 16-bit integer value at (memory) location 48.
AUTO : BOOL := 1;	Allocates dynamic memory for a symbolic Boolean variable "AUTO" with initial value = TRUE.
NAME : STRING[10];	Allocates dynamic memory to contain a string with a maximum length of 10 characters. The initial string is left to default (empty with length 0).
END_VAR;	

Examples for functions and function blocks are given in Table 3–2.

4. **Configuration elements** for memory and execution details of the code: resources, access paths and tasks.

5. **The possibility to derive** new data types, functions and function blocks, i.e., to define new ones, based on the standard ones.

Each of the five languages has its own characteristics in regional usage, syntax and possibilities for controlling the order of events. Examples of functional differences are (un-)conditional jumps to a given label, loops (FOR,

Table 3–2 Overview of the Most Used Standard Functions and Function Blocks[a]

Function Group	Examples (*function* or *function block*)
Logic Control	
—Binary logic	AND, OR, NOT, XOR, bi-stable elements
—Timers	On-delay, off-delay (TP, TON, TOF)
—Counters	Up- and/or down-counting
Data processing	
—Mathematical functions	Basic Arithmetic: ADD, SUB, MUL, ABS, DIV, MOD
	Extended Arithmetic: SQRT, trigonometric & logarithmic functions
	Comparisons: greater, smaller, equal
—Data handling	Selecting: MAX, MIN, MUX, SEL, LIMIT
—Analog signal	Type conversion, moving/shifting, formatting
	PID, integration, filtering (not as standard)
Interfacing functions	
—Input/output	BCD conversion
—Other systems	Communication
—HMI	Display, commands
Printers	Messages, reports
Mass memory	Logging

a. The functionality is offered in all five languages; however, the syntax varies.

WHILE, REPEAT), branching (IF … THEN … ELSE), array sorting, and multi-tasking.

Generally the textual languages (ST, IL) have more modeling power, which means that graphical code can be translated into textual language, but not necessarily vice versa. Only when certain rules (limitations) are followed, it is possible to automatically convert one language into another one.

- **Instruction List (IL)** is an assembler-like textual programming language, listing a sequence of simple instructions, close to the processor instruction set. IL offers bit-manipulation in a single arithmetic register. Having evolved very early for small scale applications, its usage today is less and less common.

- **Structured Text (ST)** is a procedural textual programming language that offers functionality on a higher abstraction level, comparable to

Pascal. Its flexibility makes it suitable for the definition of new (derived) function blocks or Actions in SFC.

- **Ladder Diagram (LD)** evolved from relay-logic-style wiring schematics, a format that electricians were familiar with. The graphical network thus represents a "power flow" analogous to electric power in relay systems. Especially in the U.S., LD is widely used.

- **Function Block Diagram (FBD)** is a standardized graphical representation for the "signal flow" between elements of a signal processing system. FBD is very popular in Europe as it is intuitively easy to understand and thus beneficial for communication, documentation, testing, and maintenance of control logic.

- **Sequential Function Chart (SFC)** is a representation of a stepwise "activity flow." It comprises two main language elements to order *programs* or *function blocks* in a sequence: *steps* and *transitions*, interconnected by *directed links*. As SFC elements require storage of internal state information, functions cannot be structured using SFC.

In SFC, a step represents a set of actions and is active until one or more of the subsequent transition conditions are fulfilled. In that case the step is stopped (exited) and the next step(s) is/are started; thus SFC is similar to Petri nets.

SFC *structure* can be specified either in graphical or textual form. The *functionality* (internals of the steps and transitions) must be specified in one of the four languages: LD, FBD, IL, or ST. Therefore, SFC has a special status among the five languages. It is used for batch processes and for sequences in continuous processes, such as plant startup or emergency shutdown. Additional structural elements of SFC are:

- **Branching (OR)**—When a *Step* is followed by many branches (with leading transitions) only those branches with the fulfilled transition conditions are followed.

- **Parallelism (AND)**—When a *Transition* is followed by many branches (with leading steps) all of the branches are followed in parallel.

- **Loops**—SFC allows a (conditional) backward-connection with an earlier step to enable loops in the program flow.

Small IEC 61131-3 Example

Let us consider a small example of a user defined function block (MY_CMD). Its task shall be to verify a Boolean command (MY_CMD), e.g., to open a valve. The command can be given (CMD = 1) when at least one (OR) of the two following cases is true:

- The command was automatically issued (AUTO_CMD = 1) AND the control is currently in automatic mode (AUTO_MODE = 1)

- The command was manually given (MAN_CMD = 1) AND the control is currently in manual mode (AUTO_MODE = 0) AND there is no other reason that the command should not be executed to this point in time (MAN_CMD_CHK = 0).

Figure 3–18 shows the declaration of the function block (graphical and textual) and Figure 3–19 gives four equivalent body implementations in the different languages: FBD, LD, ST and IL.

Independent of which of the languages is used during programming, they all will finally be translated into machine code to be executed on the CPU of the controller.

Trends in Control Programming Languages

For PLCs there is a clear commitment to the five standardized languages. Some companies even offer hardware-independent IEC 61131-3 programming environments to be used across different hardware platforms, which supports easy migration. Possibilities for programming new function blocks or complete programs in higher languages, such as C, C++ or Java, are on the rise.

With CPAS it is different. Even though all vendors offer one or more programming languages similar to the ones in IEC 61131-3, they often break compatibility by offering differentiating specifics to optimize usability, functionality and/or performance. This might even be reflected in a vendor-specific name for the programming language.

To name one: FOUNDATION Fieldbus (FF) uses function-block style programming but adds the possibility of easy distribution of the control functionality among all FF devices: controllers, sensors, and actuators. Attempts

3.2 Control Logic Programming

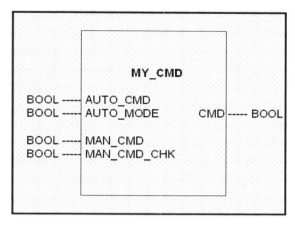

Figure 3–18 Declaration of the Function Block MY_CMD and its Variables in Graphical Form (left) and Equivalent Textual Form (right)

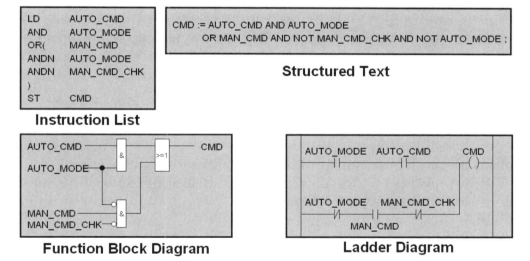

Figure 3–19 Four Equivalent Implementations of the Body of the Function Block MY_CMD in Textual (top) and Graphical (bottom) Form

to standardize *distributed* control functionalities on top of IEC 61131-3 (such as IEC 61499) have not yet made it into industrial applications (Hall, 2007).

The larger the control programs get, the more important it is not only to have good (explicit) documentation of the control code, but as well an easy to understand code (implicit documentation). This can be supported by programming guidelines; it favors graphical programming languages (SFC,

FBD) for the top level implementation and promotes concepts that are known from object orientation: hierarchical structuring, modularization, and encapsulation.

The encapsulated functionality is collected in libraries that can be reused (copied or instantiated) many times, being optimized and pre-tested beforehand. Libraries can have different granularity:

- Low level libraries offer elements for each final control device. This includes, e.g., a "temperature transmitter" including scaling, averaging, interlock logic, alarming, set-point handling, and faceplate.

- Higher level libraries might offer an "outlet control" (with sensor, PID, and actor) or a complete "centrifugal separator."

References

Hall, K.H.; Staron, R.J. and Zoitl, A. "Challenges to Industry Adoption of the IEC 61499 Standard on Event-based Function Blocks" downloaded from IEEE Xplore on 25.3.2009, 1-4244-0865-2/07.

International Electrotechnical Commission, *IEC 61131-3: Programmable controllers—Part 3: Programming languages Edition 2.0.* Geneva: IEC 2003. ISBN: 2-8318-6653-7.

John, K.H. and Tiegelkamp, M. "IEC 61131-3: Programming Industrial Automation Systems, English Edition", Heidelberg: Springer-Verlag, 2001. ISBN: 3-540-67752-6.

3.3 CPAS Functional Units

Christopher Ganz

Introduction

Automatic control of an industrial plant is, at first sight, quite complex. Typical plants have a control system installed that comprises several hundreds to tens of thousands of measured and controlled variables (I/O points). Depending on the plant type, binary and analog signals, and input and output signals, are all in use. Typically, there are more binary than analog signals, and more input than output signals.

To come up with a functioning control application for a plant, a well structured approach is necessary, one that will fulfill several requirements:

- **Specification adherence**—The resulting application must fulfill the specification defined in the plant function. In some cases this may also mean compliance with norms and regulations, or even certification.

- **Operability**—The application must be operable not only in the fully automatic use case, but also in any situation in plant operation, such as maintenance, recovery procedures, etc.

- **Maintainability**—It must be possible to maintain the application over the lifetime of the plant, which is normally longer than any software life cycle.

Designing the application functionality in one interwoven block of code will not result in the desired structure. As in software development for traditional IT applications, a functional decomposition is required. In the case of automation, the approach presented follows the functionality of the plant and the control hierarchy.

A well structured definition of the automation functionality is one key step towards achieving the goals mentioned above. In a second step we will also look into some implementational aspects. Again, looking into software development practices, we derive some of the proposed concepts from object-oriented programming. This technique is widely used and can easily be applied to automation software development.

Example: Heated Liquid Tank

In this chapter, we will repeatedly refer to the following example, which is illustrated in Figure 3–20.

To operate, a process requires a liquid to be at a defined temperature. This liquid is heated in a tank. The tank comprises the following components:

- Tank, with level indicators low, normal, and high
- Heater, which heats the liquid. It is located below the "low" indicator, i.e., the heater must not be on below that limit
- Pump, which feeds the tank with cool liquid
- Drain valve, which empties the tank completely when open
- Pipe to the process where the warm liquid is consumed. The use of the liquid is controlled by the consuming process

We will refer to this example as the "heated liquid tank".

Automation Design

The goal of an automation system is to operate a plant, or parts of it, automatically. Although this may seem a trivial observation, it needs some further explanation:

- **Plant**—The complete plant is hardly ever run automatically. There are always processes that are out of the scope of automation (e.g., material supply), or different sections of the plant are automated, but not its entirety. Different automation systems may be active in warehouse management, production, and packaging, just to name a few examples. These plant areas are coordinated by operator interaction.

3.3 CPAS Functional Units

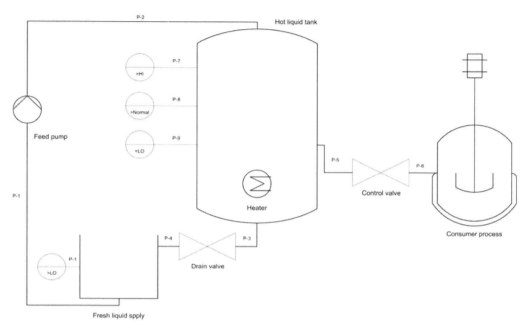

Figure 3–20   Heated Liquid Tank System

- **Operate**—Even if a plant is fully automated, for the most part only its main operational modes are run automatically. A plant may be started and stopped automatically, it may produce automatically, but to cover all possible modes of operation by automatic functions exceeds the capability of an automation system. Maintenance situations are normally not automated (or not to the full extent), and require manual interaction with the maintenance personnel. Abnormal situations are not covered either, in many cases. If a device fails in operation, the plant either trips, or requires the attention of the operator to complete an operation, or safely shuts down the equipment in question.

- **Automatically**—In most plants, the operator is still the one responsible for proper plant operation. Some operational concepts require an operator to manually release the next production step, or to trigger operations based on his or her observations or on operating procedures defined outside the automation system. Instructions like "make sure no person is near the equipment when starting up" are frequently checked by personnel, and need to be confirmed to the automation system. Even in a fully automated plant, some manual interaction may be required.

The design of the automation system needs to take all this into consideration. This is normally done by allowing manual interaction with the automation system on different levels of the control hierarchy: manually operating a device, starting a function, tripping the plant, etc.

Since manual interaction always carries the risk of human error, measures need to be taken in the automation system to prevent dangerous interaction, and to support the operator even in manual tasks.

These boundary conditions need to be considered when designing a safe and reliable automation solution.

Control Hierarchy

Automation systems are typically designed hierarchically (Figure 3–21). From the direct interaction with a device in the field up to the automatically operated plant, a whole hierarchy of devices, process functions, and subsystems helps to allocate the automation function where the appropriate information is available.

Manual Interaction

To allow for manual interaction as mentioned before, direct interaction with a device is normally possible. This may be done directly from a local panel on the device, from the switchgear cabinet, or by directly manipulating a process variable in the control system.

In case of maintenance, diagnostic information on the device is normally available on a similar level, e.g., on the device's operator panel, or on a faceplate of the process variable. This information is not only available for actuators, but also for measured values. Alarm states, disturbance indicators, or detailed status information can be read from the device display or a faceplate.

Working on this level of interaction is normally only available to maintenance personnel, and is not required during normal operation of a plant.

The lowest level of automation in a system is controlling a single device, or a loop. The applications are dominated by two loop families:

- **Binary control**—a device (pump, motor, etc.) is switched on and off, or is opened or closed in the case of a valve.

- **Analog control**—a process variable is controlled to a set point by means of a (typically) analog device that accepts a continuous variable as its input (variable speed drive, control valve, etc.).

3.3 CPAS Functional Units

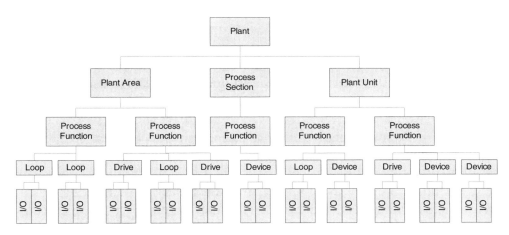

Figure 3–21   Plant Automation Hierarchy

Apart from closing the loop (for example, binary control: switching the motor off when the valve has reached the end position; analog control: PID loop), device control may include some extended functionality:

- **Interlock signals**—prevent the device from operating in undesired conditions
- **Diagnostics**—detect improper behavior of the device
- **Operator interface and alarming**—representation of the device at the operator station

Device control is normally done using standard libraries that provide all the desired functionality required. These libraries are normally industry-specific.

If the control of a plant is left to individual loops, the level of automation stays very low, and the plant still requires a lot of manual interaction with the operator in order to properly start up or shut down. The coordination of individual loops is the main component of a fully automated plant.

To structure the overall plant function, it is hierarchically divided into a number of functions which comprise a limited number of loops and measurements each, and perform one specific process function. These functions are then coordinated on a higher level, and may be combined into more complex functions themselves, comprising sub-systems, plant areas, or even plant units.

The proper functional decomposition of the automation software into manageable functions is at the core of every large automation problem.

Functional Design

The first step in the design is the functional decomposition. If you approach the process looking at the installed hardware, you will most easily identify systems and sub-systems as process equipment grouped together. If these systems are supplied separately (gas compressor, fuel oil supply platform), this approach is suitable.

This approach is actually required when the control of that specific sub-system is to be done locally, i.e., close to or within the sub-system.

The size of a process section does not matter in the first step; a functional decomposition is required in the next step. If you see in this decomposition that it is reasonable to further divide the section into sub-systems, you are free to do so.

Consider the fuel oil supply platform as an example. Clearly this is one sub-system, since it is mounted on one platform, it has clear interfaces from the process point of view (oil in and oil out), and it contains a relatively small number of drives and pumps to be controlled. Its function is easily identified: provide fuel oil. The open question is, how much?

1. Identify a system according to the process structure.

2. Clearly identify all interfaces between the system and the control system (measurements and drives). Each I/O signal in the plant should belong to exactly one section, system, or sub-system. These signals then become part of that sub-system.

3. Identify the system's function from a process point of view (example: "supply fuel oil to the plant").

4. Identify the system's function from an automation point of view.

5. If you end up with too many functions in step 3 or too many interfaces in step 2, consider defining sub-systems recursively.

Once the scope of the functions has been defined, the next step is to design the control functionality. Many functions of a process have a simple control interface: the function can be either on or off. These are the two states of the function, and it can be switched between these. It should be kept in mind that a more complex process function is not simply switched on and off, but may need to be brought from the "off" state to the "on" state in a more complex sequence, i.e., starting up the process, and similarly controlled when the process is stopped.

Analog functions have an additional parameter in their "on" state that controls their function: an analog set point. These functions normally have off and on states, and can be started and stopped, but when in operation, a desired value can be set in addition.

Even a function with only two (steady) states (on/off, stopped/running) may have more internal states that need to be considered. As we have seen before, a function may be brought from the "off" to the "on" state with more complex automation procedures. It will therefore be in a "starting" transient state for a considerable amount of time (and also in a "stopping" state when being switched off). If the process is interrupted during one of these transient states, there are two things that may happen: it may be in an "undefined" state, with some of the equipment in the "off" position and other in the "on," or if an undefined state is not allowed or it encounters unexpected process behavior, it may trip, i.e. it is brought to a safe state from whatever state it is in. In many examples, the "tripped" state is not identical to the "off" state. The function therefore needs to be brought to the "off" state in a controlled manner.

Functions with more than two steady states are less frequently observed. The extension from two to more states is slightly more complex: whereas in the two-state case the function is always brought from some initial state to its opposite, in multi-state, and depending on the starting state, functions may require different automation schemes when brought to the same target state.

Designing the automation function requires looking at the steady states (what needs to be done when the function is on, e.g., level control, counting, etc.), and what needs to be done when the function is brought from one state to another.

In binary as well as in analog control, in order to close the loop on any level, process feedback needs to be provided. Since only the function in question can decide whether it is properly operational, function feedback needs to be created for any function that accepts commands or set points. This feedback reflects the state of the function with respect to the given commands or set points. The function feedback is then used by neighboring or supervising functions to detect whether a command was executed properly.

The typical use of function feedback is in sequence control (next section), where plant areas are started in a defined sequence. The next function usually can only be started when the previous function is properly in operation.

In addition to information as to whether the function is properly started or stopped, functions may provide information about disturbances. This information is normally provided to the operator in the form of alarms. However, since neighboring or supervising functions cannot access alarm information, diagnostic information is typically also provided by a function feedback signal. In many cases, a simple "disturbed" signal is sufficient; in more complex automation systems, the information may be more detailed and may indicate a number of possible disturbances.

Starting or stopping a plant or plant area is frequently done stepwise, with process functions starting as soon as the preceding function is properly running (Figure 3–22). Instead of manually programming the interdependence between functions, this sequential procedure is coordinated by means of the IEC 61131-3 sequential function chart programming language. It allows the programming of a finite state machine, where state transitions may only occur when the transition conditions are fulfilled.

Sequencers are typically used to start or stop a function, plant area, or plant. The controlled function has two states (on and off), where the state transition initiated by an "on" or "off" command is done by means of the sequencer.

Since a plant may be operated in manual mode under some circumstances (maintenance, failure recovery, etc.), the initial state of a function may not be known initially. Its devices may be in a set of states that do not conform to either the "on" or the "off" state. In that case, an initializing function may be used to bring the function to a safe and defined state. Frequently, the off-sequencer can be used to perform the initialization.

If a function has several stable states, the proper automation function may be implemented with one sequencer per allowed state transition.

In addition to the manual interaction of the operator with individual devices in a function, the sequencer itself may be operated manually. Options such as manually stepping through the sequence (with indication of transition readiness) may be available on the operator station.

Functional Variations and Re-use

One important aspect of a properly designed hierarchical functional decomposition, as shown before, is to achieve robustness against changes, and to increase the possibility of re-use of software components.

Re-use is possible on different levels and can mean pure copying of a previous version, encapsulated function code, using templates, etc. Re-use can mean using a piece of software from an earlier project, or in a different context. We can see various scopes of re-use:

3.3 CPAS Functional Units

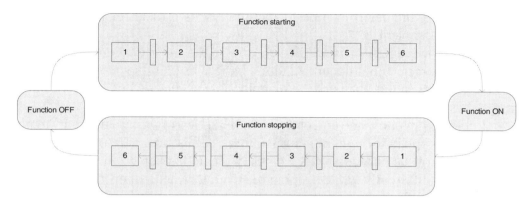

Figure 3–22 State Diagram Including Start and Stop Sequencers

- **Standard procedures**—Functions which are used in numerous places, independent of the related process or function. They are limited in complexity, but wider in applicability. These functions are typically maintained as libraries. In many cases, they are common within an industry.

- **Variants**—Functionality which can be available as different variants. Modules with the same functionality, but with different internals, can be exchanged, or different implementations of the same process function (e.g., hydraulic or electrical) may be chosen in a project. The re-use benefit in this case comes with the stability of the enclosing application, which remains unchanged for any variant used.

- **Functional units/sub-systems**—Functions are common in different types of plants, e.g., a compressor of a certain type, a digester, a mill.

The re-use potential of a reusable component is influenced by two factors:

- **Coverage**—How much of the project is covered by one standard module.

- **Re-use Frequency**—How many times a standard module can be used.

High coverage usually means a lower re-use frequency, since the probability is higher that a current project differs in some detail from the standard module and needs some modifications. On the other hand, the highest benefit is achieved with a high coverage and a high re-use frequency. Since this is very rarely possible, the determination of module size is critical for the success of the modular approach.

Change management is very often one of the most challenging tasks in automation design. Input documentation changes as plant development moves on, and changes in the design of the plant have an immediate effect on the automation system.

Looking beyond one project at a whole series of installations, in many industries the plant design remains similar, but is adapted to the specific requirements of the particular job. This may mean that some functions may be implemented differently while keeping the overall structure constant.

One of the key challenges to an automation application with regard to plant modifications is to track the influence of the change, and then adapt all components to accommodate the changed functionality.

Proper functional decomposition and encapsulation is a key technique applied to reduce the impact of a modification. If other functions do not rely on internal signals or variables of a sub-system, but on information provided by the encapsulated sub-system function, the change can be encapsulated as well. Since other parts of the software solely rely on a piece of information that has been derived from measurements, the effect of a change can be kept local by newly creating that piece of information in the context of the changed function.

If, for example, the information on whether a hydraulic system is functioning properly is derived from the hydraulic pressure level, this measurement is lacking if the mechanical pressure sensor is replaced by an electrical implementation. If other functions do not rely on the "hydraulic pressure high" signal, but only on the "system on" status signal derived from the pressure level signal, the same status signal can be derived from the status of the electrical system. In that case, the "system on" status signal may be a breaker position, or even a voltage measurement.

Changing the function implementation in the process then does have an effect on the automation of the sub-system in question (which may be completely different from case to case), but the enclosing functionality of the automation system remains unchanged.

While object variants kept the application surrounding an object stable and exchanged its functionality, the opposite is possible as well. In case of a re-used sub-system, it may appear in varying contexts; hence the enclosing logic varies while the object's function stays constant.

Combining a set of previously defined encapsulated functions in a project application helps the engineer to focus on the integration work rather than the details of the object's function. The level of abstraction of the engineering task can be increased by concentrating on the system integration aspect of engineering rather than on application level programming.

Both approaches, either keeping the plant structure constant and varying the implementation of a process section, or combining pre-defined functions to form a project-specific application, rely heavily on one key component of the functional implementation of the automation system: stable interfaces.

If interfaces between functions are defined early in a project, or can even be standardized throughout a whole family of plant types, variations in projects can more easily be allocated, tested in the context required, and commissioned.

When designing the runtime interfaces of a function, the information requirements define the framework of what a function needs to know about its surrounding processes, and what commands it must accept from supervisory control. Complementary to that, we need to define which information about a function is required by its environment.

A good classification of the information exchanged by a function and its environment is shown in Table 3–3.

Please note that the information exchanged between two functions should reflect the level of abstraction in that functional context, and not assume the internal behavior of one of the functions. In the heated tank example, the information as to whether hot liquid is available for the consuming process should not query the tank level directly. Instead, the tank function should provide the "hot liquid available" signal (that is composed of the liquid level and the temperature measurement). In a case where there is no tank, but only a continuous flow heater, the "hot liquid available" signal is still viable, but not in the form of the liquid level; instead maybe derived from a flow measurement. The function of the consuming process therefore does not need to be changed when the feeding function is modified, and both variants provide the same information, but not the same variable.

In addition to the dynamic information an object requires for its automation tasks, it may also need some static information that is available in the engineering or configuration stage only. This information may be given in the form of parameters that may only be changed through the engineering tool and not necessarily supplied continuously at runtime.

Examples for parameters set during the object's configuration are controller gain parameters, plant constants, measurement ranges, etc. (Figure 3–23). These parameters are normally set at the engineering stage, and only changed during commissioning or maintenance.

Table 3–3 Information Exchange

Commands	Input signals that change the internal state of the system. These may be on/off commands, set points, or safety shutdown signals. In the same category we find mode switch request signals (e.g., manual operation) that are indeed commands that change the internal state of a function, but are very often treated separately.
Feedback	Output signals that indicate what state the function is in. As a rule, for each command there should be feedback that indicates that the desired state has been reached. In the case of a set-point signal, the corresponding controlled value should be fed back. Similar to the mode switch commands, the current mode of the function should be indicated in a feedback signal.
Environment Information	It is to be noted that from within a function only process signals from that particular function should directly be accessed on the I/O variable. If information is required from another function, it needs to be available as an input signal to the function. In this category we find interlock/release signals that prevent a function from being started.
Information Provided	Similar to the information required from other functions, a function needs to provide information to other functions at its interface.
Hardware I/O	Functions that are executed in one device, e.g., a single drive or actuator, also comprise the corresponding interfaces towards the I/O signals (device feedback inputs and command outputs).

Similar to the parameters that are only set at specific instances by the specialists, an object may provide diagnostic information to maintenance personnel. And here as well, it is not required that this information is continuously transmitted throughout the system. It may be accessed only in case of malfunction or to debug the module in question.

From a functional perspective, anything that prevents a function from properly executing is classified as a disturbance. The most simple disturbance monitoring is therefore common to all functions: A command to reach a certain state has been received, but the function does not arrive at the desired state within a reasonable time. Since we have seen that all functions require command input that triggers state transitions, as well as state feedback signals that indicate whether the state has been reached, this generic supervisory function can always be implemented. More detailed analysis of the failed function may need to be implemented, depending on the desired functionality.

3.3 CPAS Functional Units

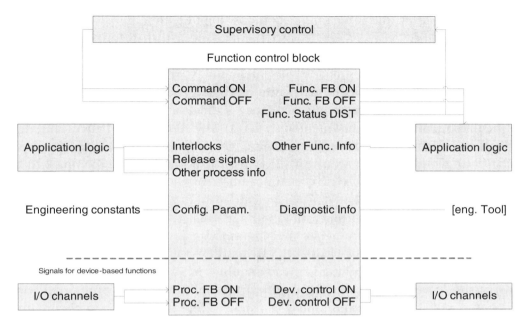

Figure 3–23   Function Interfaces

A special case of diagnostic signals is the variables that are used as alarms. These variables have a normal operation range, and an alarm state, i.e., if the variable exceeds the alarm limits set (or in case of a binary variable, switches from the normal state to the alarm state), an alarm is raised and displayed on the operator station to be investigated and acknowledged. Depending on the alarming philosophy, it may be desirable for the function to raise a summary alarm on the function state, which allows the operator then to dig into the alarm list for that specific function to diagnose the reason for the failure of the function.

Functionality

The functionality of a module defines how it handles its subordinate automation structures, or its process section. The module functionality is developed based on the desired behavior of the controlled process.

The main functionality defines the behavior of the module, assuming all process equipment is working properly. Each of the provided functions must be available to an external command, and each function must return a status value as to whether the function was completed successfully.

In most cases, the functional decomposition of a plant results in one function per module ("cool," "heat," "provide air," etc.); it therefore provides the commands to switch that function on and off (and returns the corresponding feedback signals).

More advanced automation systems not only handle normal operation, but also provide a means to circumvent process disturbances automatically. The typical application trips the function in a safe way. Other disturbance handling functions are failover to redundant devices, or switching to a backup function. Handling disturbances in the automation code is normally more complex than the undisturbed function, and is at most done for the most critical or foreseeable failures.

In addition to normal operation and disturbance handling, it may be reasonable to add commissioning or test support to a process function. Very frequently, commissioning is assumed to be working by directly interacting with process variables either by manually controlling individual devices or by forcing controller variables using the engineering tool's commissioning functions. This may result in dangerous situations, since safety interlocks can be circumvented that way.

A safer way is to provide specific functionality that supports commissioning of the plant. If commissioning requires well defined experiments to tune controllers or set points, these experiments may be provided in additional commissioning code that can be accessed from the operator station rather than from the engineering tool. To reduce the controller load during normal operation, commissioning code can be removed from the application when the plant is operational, or it can be disabled. To properly remove the code, encapsulation and modularization as previously described are essential in order not to damage the rest of the application in this operation.

Example: Heated Liquid Tank

Let's return to the heated liquid tank example, and try to apply the concepts we have presented. First, let's look at the heated liquid tank in its process environment (Figure 3–20). The tank receives liquid from the fresh liquid supply. Excess liquid is fed back to the fresh liquid supply as well. The heated liquid is provided to the consuming process by a control valve.

3.3 CPAS Functional Units 205

Functional Decomposition

In this chapter, we will repeatedly refer to the following example, which is illustrated in Figure 3–20, the heated liquid tank is embedded in its surrounding process: There is a fresh liquid supply system (depicted as a single tank), and the process that consumes the heated liquid, including all actuators required.

The first step in the functional decomposition is to define the function of the equipment in question. The overall function of the complete system shown is to produce the result of the consumer process, whatever it may be. The function of the tank sub-system is then defined as "provide heated liquid to the consumer process."

The overall function "produce finished product" is therefore divided into the three functions:

- Provide fresh liquid to the heated liquid tank
- Provide heated liquid to the consumer process
- Process product

A functional diagram may therefore look like the one depicted in Figure 3–24. It still lacks the interfaces between the functions, which will be developed in the next section.

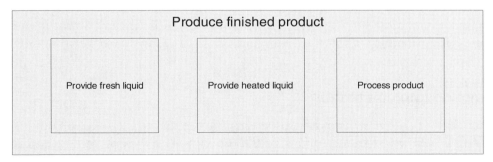

Figure 3–24 Function Diagram of Tank Example

State Encapsulation and Interfaces

Let's look at the interfaces required to run the heated liquid tank function in the context of the system depicted in Figure 3–24. The first step is to define the internal states of the function. The phrase defining the function, "provide heated liquid" can only have two states: yes, the function does provide heated liquid, or no, it does not. These two states can be assumed to correspond to an on-state, and an off-state of the function.

To define the interfaces, we will follow the classification as described in Table 3–4.

Device Allocation

Looking at the diagram in Figure 3–20, we now need to decide which devices belong to which function.

In order to provide the desired functionality, the function "provide heated liquid" needs to be able to:

- Control the liquid level in the tank. Therefore, the feed pump, the drain valve, and the level measurements need to be part of the function.

- Heat the liquid. The heater therefore also needs to be part of the function.

Since the position of the control valve depends on the control algorithm of the function "process product," assigning it to the tank control function would add to the complexity of the interface, since the "process product" function would then need to tell the tank control how much liquid it requires. If the control valve is assigned to the "process product" function, it can control it within its own functionality without exchanging information with another function.

Function Implementation

After defining the interfaces and allocating the devices to the function, all information is available to actually design the function, i.e., the controller that keeps the level and the temperature within the heated liquid tank constant. Since there are only three binary variables that indicate the tank level, the level control is bound to be simple. A two-state controller is the easiest implementation, switching the pump on when the level is low, and switching the pump off when the level is high. The heater can be turned on as soon as there is enough liquid in the tank (low limit reached), and it needs to be shut down if the liquid level stays below the low limit for some time.

3.3 CPAS Functional Units

Table 3–4 Interface Definition for the Heated Liquid Tank Function

Commands	Since the function has two states, on and off, there are two commands: turn on and turn off, or, since providing the heated liquid is a continuous action, start up and stop.
Feedback	Corresponding to the two commands, the function returns two states: on and off, where "on" means the function does provide heated liquid, and "off" means the function does not.
Environment Information	In order to properly work, the function only needs one bit of information from its environment. To properly start, the function needs to know whether there is fresh liquid in the supply tank. Hence the preceding function "provide fresh liquid" needs to export the information as to whether fresh liquid is available.
Information Provided	Other than the information as to whether heated liquid is available (status feedback "on"), the following function "process product" does not require additional information.

Function Diagnostic

The interfaces defined so far are sufficient to automate the process in an undisturbed environment. However, to properly cope with disturbances, more information is required.

In addition to a generic "discrepancy" function that indicates the failure to reach the desired state, the heated liquid tank function could provide information such as failure of any of the devices, or low/high liquid level although the function is running properly.

The diagnostic information can be made available to the enclosing logic to react to a failure of the process. However, since most failures need to be taken care of by maintenance personnel, this information is normally brought to the operator station in the form of alarms.

Conclusions

Re-use of automation functions is hardly ever considered early in a project. Project requirements are met as efficiently as possible, while re-use is left to the people working on the next project. If they can easily use what was done earlier, the better it will be.

Consequently introducing the functional unit approach presented in this section, builds re-use capability into automation software from the beginning. Will a project that incorporates this approach therefore be punished in favor of the projects to follow? We have shown that the structured approach not only favors re-useability of the engineered solution, but also makes it more stable within a project. The application becomes less sensitive towards late changes in a process area, and is easier to test and commission. Furthermore, if the methodology is introduced in an organization, applications will be easier to maintain, since they follow a common design scheme.

Once introduced, an organization can start building libraries of automation solutions that further reduce engineering effort on new applications. If these libraries are even maintained by feeding back experience from plants, a continuous increase in quality can be achieved.

References

Johannesson, G., *Object-oriented Process Automation with SattLine*, Studentlitteratur, Lund, 1994.

Ganz, C. and Layes, M., *Modular turbine control software: A software architecture for the ABB gas turbine family control system*, Second International Workshop on Development and Evolution of Software Architectures for Product Families, Las Palmas, 1998.

CHAPTER 4

4.1 Alarms and Events

Martin Hollender

Introduction

The intuitive interpretation of the word alarm is that it is something important that requires attention. Unfortunately, in many real plants the situation is quite different, with thousands of alarms generated every day, most of them having no value at all for the operators.

In process automation, the terms alarm and event are not 100% clearly defined. The OPC Foundation's Alarm and Event specification (OPC AE) is the most commonly used standard to communicate alarms and events between different modules. It defines an alarm as a special case of a condition, one which is deemed to be abnormal and requiring special attention. Alarms need to be acknowledged by the operator. Areas of interest include safety limits of equipment, event detection and abnormal situations. In addition to operators, other client applications may collect and record alarm and event information for subsequent auditing or comparison with other historical data (OPC AE 1.1 specification).

OPC AE defines an event as a detectable significant occurrence which may or may not be associated with a condition. Events cannot be acknowledged by the operators. An event is a single point in time whereas an alarm has a start and an end (often called Return To Normal [RTN]). Typical events that can be associated with an alarm are start, end and acknowledge events (Figure 4–1).

Guidelines like EEMUA (Engineering Equipment & Materials Users' Association) 191 (see Section 6.2) have emphasized the fact that the ability of humans to process messages is limited. From that point of view, an alarm is everything that is brought to the attention of the operator, be it as an entry in a list, a sound in the control room or a color change in a graphic display.

4.1 *Alarms and Events* 211

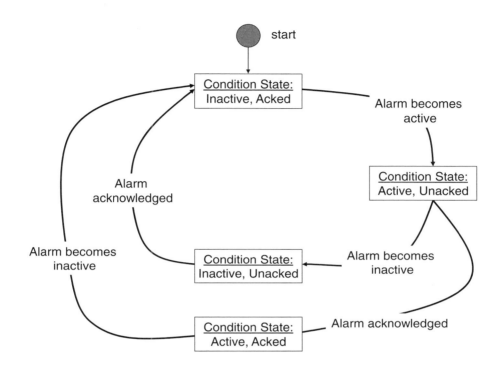

Figure 4–1   State Diagram Showing Transitions Between Alarm States

Figure 4–2 shows the typical layers involved with alarms. The most common alarm source are function blocks in application code running on the automation controllers. OPC AE defines three different event types:

- Condition-related events (typically associated with alarms) represent state transitions. An example is a pressure exceeding a certain threshold. The alarm activates when the threshold is exceeded and returns to normal as soon as the pressure has fallen below the threshold again.

- Tracking-related events can represent single points in time. An example is the message that an operator has opened a valve.

- Simple events represent everything else.

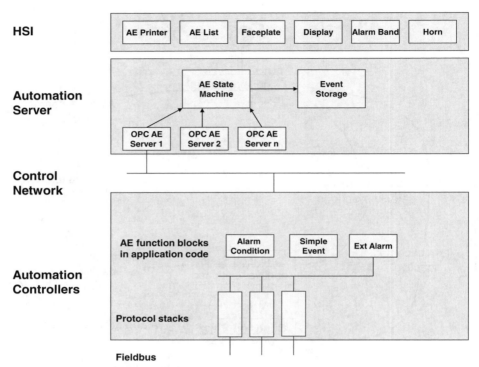

Figure 4–2   Alarm System Architecture

Alarm and Event Sources

Typical alarm and event sources in a CPAS are:

- Automation controllers (e.g., temperature too high)
- Hardware units like fieldbus devices
- Software applications such as asset monitors supervising equipment
- The system itself (disk full, service unavailable, etc.)
- Sometimes process alarms (like temperature too high) are implemented in a level above the automation controllers. Two different motivations exist:
    - The automation controllers might not have sufficient (processing) resources to implement the alarm logic or do not offer alarm functionality at all.

4.1 Alarms and Events

- o The separation of control logic and alarm logic can result in more transparent alarm logic, because otherwise the operators might be forced to go down to the control logic to understand the details of the alarm logic.

Usually the interface to all those different sources is OPC AE. A larger plant has a multitude of different OPC AE servers from different vendors with different philosophies. It is the task of the CPAS to harmonize heterogeneous alarm sources and ensure that each alarm is routed to the recipients who need to know about the alarm.

Process alarms are often generated in automation controllers. Typically process variables like flow, pressure, temperature and level are compared with high and low thresholds. Once a threshold has been exceeded an alarm is generated. For a single process variable, multiple alarms can be configured such as LO (low), LOLO, LOLOLO, HI (high), HIHI, HIHIHI and rate of change alarms.

In Figure 4–3, the parameter FilterTime determines for how long the signal must deviate before a change is considered to have taken place, because otherwise every little glitch would cause an alarm. So-called chattering alarms can create a lot of flurry in the alarm lists presented to the operators. It is therefore important to configure adequate filter parameters.

Alarm Indicators

Historically, alarms were indicated on special annunciator panels. Modern CPAS offer several software-based alarm indicators, but the ability to use special hardware like printers, horns or lamps remains important. Alarm indicators include:

- Alarm and event lists
- Alarm bands—An alarm band summarizes an alarm list and provides a link to the corresponding alarm list display. The number on an alarm band represents the number of currently unacknowledged alarms in the corresponding list. The color of the alarm band shows the highest priority alarm active at the moment.
- Alarm sequence bars—The alarm sequence bar is a status display, where the most recent alarms are displayed horizontally right to left in

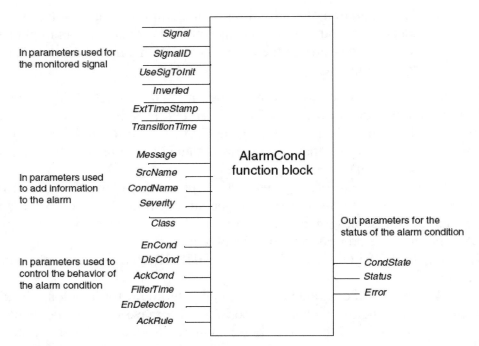

Figure 4–3   AlarmCond Function Block as Alarm Source

the order of their occurrence. The alarm sequence bar allows calling up the context menu for the alarm, acknowledging the alarm and viewing further details.

- Alarm indicators on display elements

- Spatial dedicated alarm displays—The old alarm annunciator panels had the advantage that an alarm always appears at the same place. As Woods points out (Woods, 1995), spatial dedication supports preattentive reference. In modern screen-based systems, spatial dedication is sometimes lost. It is therefore very important to provide an overview display (e.g., on a large screen) that contains spatially dedicated continuously visible indicators for the most important alarms (EPRI, 2005).

- Alarm printers—The traditional way of logging alarms was printing them on paper. In some control rooms the noise generated by the matrix printer was an indication of how well (or not) the process is running. However alarm printers have some severe disadvantages. A constant supply of consumables like paper and ink needs to be ensured. The

4.1 Alarms and Events

(sometimes very large) piles of paper need to be archived. The printers need to be regularly maintained and constantly monitored. As a result, alarm printers are currently being replaced by databases.

- Alarm historians—Alarm historians are usually part of the Information Management module (see Section 5.5). Alarms and events are stored in a database and can be retrieved for all kinds of reports. Alarm historians are the basis for a detailed alarm analysis.

- Audible and external alarms—Active unacknowledged alarms can generate a sound on the local workstation (e.g., play an mp3 file). Such alarms can be configured to be silenced either from the local workstation only or globally, from all workstations. If special hardware like horns or lamps is to be used to indicate the alarm, this can be done with the external alarm function. It routes the alarm signal to specially configured I/O points.

- Email—Detailed alarm information is sent to a configurable email address. Typically email is used for maintenance-related tasks which don't need an immediate response. For example, an email can be used to notify an external service center that a device needs closer inspection. Email can also be used to start workflows, e.g., insert tasks into ToDo lists.

- Notification to external systems, for example Computerized Maintenance Management Systems (CMMS)—Many plants use CMMS (see Section 5.4 for details) as a basis of their maintenance work. No maintenance action, be it inspection, repair or reconfiguration can be done without a work order in the CMMS. Therefore maintenance-related alarms requiring a maintenance action can automatically create a work order in the CMMS to start a corresponding workflow.

- SMS (Short Message Service)—Some highly automated processes like district heating plants don't require a 24 hours a day, 7 days a week staffed central control room. The recipients for alarms (operators and/or maintenance staff) are mobile, occupied with other tasks elsewhere or on on-call duty. In such cases, it is important that the CPAS can send alarms as SMS messages. If an SMS message is not acknowledged, it can automatically be resent to another operator until the SMS is acknowledged to make sure the alarm is being taken care of.

Interacting with Alarms

The standard workflow with alarms can be explained with the help of the following example:

1. **Activation**—For example, the level in a boiler rises above an alarm threshold. If no action is taken, the liquid might flow over.

2. **Acknowledgment**—The operator signals that he or she has noticed the alarm and takes responsibility for fixing the problem. The acknowledgment becomes immediately visible in all other workstations where the alarm is being displayed (unacknowledged alarms are often blinking). In the case of high time pressure, the operator might skip this step

3. **Operator action to fix the problem**—The operator might call up a faceplate to open a valve or might call a field operator to manually open a valve. If the reason for the alarm is unclear, the operator first needs to diagnose the problem. Often the operator cannot fix the problem but needs to forward the problem to the maintenance department.

4. **Return to normal**—The level falls below the alarm threshold and the alarm deactivates.

If the alarm is safety related (e.g., if the liquid is explosive as in refineries), a safety instrumented system (SIS) threshold needs to automatically make sure that the level in the boiler never gets too high, if necessary by automatically shutting down the plant. A plant may not be designed in such a way that safety relies on an operator reacting properly to a safety-related alarm (IEC 61508 and IEC 61511).

Context Menu

When an alarm appears in an alarm and event list, typical questions concerning this alarm are:

- Why was this alarm activated?
- What does the related trend display look like?
- Have similar problems occurred in the past?

4.1   Alarms and Events 217

- When was this device last maintained?
- Where can I find documentation to understand the problem and the device better?
- Which faceplate can be used to fix the problem?

A CPAS needs to make such information available just by clicking on the alarm and activating the context menu.

Shelving

Often alarms grab operator attention by activating the horn and adding lots of entries to the alarm list, which might make no sense in the current situation, e.g., because the equipment is currently under maintenance or because the equipment is known to be faulty but cannot be repaired in the near future. The shelving facility of an alarm system allow the operators to shelve alarms, which means putting them temporarily aside. The alarm shelf needs to be continuously reviewed to make sure that no alarm is forgotten. The CPAS can support a systematic management of change of shelved alarms by providing periodic reminders and approval mechanisms.

Disable

Disabling an alarm means to turn it off. This can make sense, because sometimes alarms have no value at all for the operators. For example, a faulty device might generate thousands of alarms per day, but cannot be replaced until the next maintenance shutdown in a few months. As in the previous example, it is very important that the alarm is re-enabled once the device has been replaced.

Often the value of alarms changes over time, e.g., when additional process parts are activated or if the process is run differently.

Another problem is that some alarms in the system might be very well engineered, with lots of thought behind them, whereas other alarms were configured just by default. It is therefore important to have good documentation available showing the rationale behind every configured alarm and a sound change management system.

In some CPAS, disabled alarms are still generated and available for logging, whereas in others disabled alarms are treated as non-existing.

As with shelved alarms, the list of all disabled alarms needs to be closely monitored and reviewed to make sure that important alarms are not disabled. Proper authorization and change control are required to make sure only personnel with the required knowledge and authorization can disable alarms.

Comments

A comment facility allows operators to attach comments to alarms. These comments are available as long as the alarm is active and can be logged to the alarm historian. Later on all comments that were attached to an alarm can be retrieved and might help to diagnose the current problem. Alarm comments also help to build up explicit knowledge about what is going on in the plant and are an important basis of operator empowerment and continuous improvement.

Alarm System Features

Hiding

Hidden alarms have a reduced visibility. They don't grab the operator's attention in the first place, but hidden alarms are still available if explicitly requested. For a detailed and in-depth diagnosis, the operator can view all alarms, including the hidden alarms.

Alarm hiding is implemented transparently in the alarm system layer above the controller. Typical hiding rules include:

- If A is active, don't show B or C.
- If state = startup don't show D, E, F, G.

Alarm lists can be configured to exclude hidden alarms. By clicking one button, hidden alarms can be included again. Figure 4–4 shows how alarm disabling, hiding and shelving cooperate to tame the alarm flood.

Filtering

Modern CPASs offer powerful filtering mechanisms. Some older DCSs had only a single alarm list showing all alarms in the system. The much higher

4.1 *Alarms and Events*

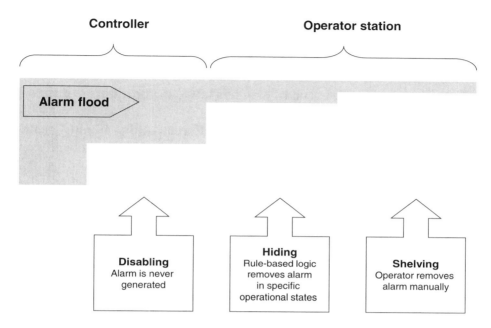

Figure 4–4   Removing Alarms at Different Stages

alarm rates and increased diagnostic capabilities of modern systems make such an approach impracticable. A CPAS allows configuring specific alarm lists for special purposes. The most important filter criteria include:

- Target group (operation, maintenance, system)
- Priority (urgency and importance). Alarms during a process upset may require different operator focus than alarms during normal, smooth operation. Alarms used to fine-tune and optimize the process make no sense during a process upset, when it is important to focus on the most important issues first.
- Process area. Often different operators are responsible for different process areas. Their main focus is the alarms of their area, but alarms of other areas might be relevant also if causal relations with those areas exist.
- Many other special-purpose alarm lists, based on entries in vendor-specific attributes (more on these later in the chapter).

Priorities

Each alarm and event has an assigned priority value which indicates its urgency and importance. Well-engineered priorities help operators to focus on the most important issues first. In many existing plants, alarm priorities are not very well assigned or not used at all. In such a case it is left to the operators to decide if an alarm is important or not.

Usually priority values are mapped to different colors. In most cases the priority assignment to an alarm is static, though it would be theoretically possible to calculate situation dependent priority values.

The OPC AE standard uses the term severity as a synonym and allows the use of priority values between 1 and 1000. This is unfortunate, because most humans cannot distinguish among so many priority levels. The EEMUA 191 guideline recommends the use of only three different main priorities (high, medium, and low). Critical priority can be assigned to a few alarms as an exception.

Alarm priorities are defined at the source and therefore it is not always possible to ensure a consistent, system-wide priority assignment. For example, one part of the system (e.g., a legacy system) might use the priorities 1, 2 and 3 for high/medium/low whereas another part of the system might have the mapping shown in Table 4–1.

Such inconsistencies are confusing at best, if not outright dangerous. An operator used to the assignment in the second part might misinterpret a high-priority alarm coming from the first part as low priority.

A CPAS that integrates alarms coming from different sub-systems must be able to harmonize the priority assignment. In the example above, the CPAS might remap the legacy priorities 1, 2 and 3 to the CPAS priorities 900, 700 and 500.

Systematic Definition of Alarm Priorities

Priorities should be defined in advance during plant engineering to indicate which alarms are more urgent and important than others. Usually there is a tendency to claim that most of the alarms are of high priority. EEMUA 191 recommends a distribution of 5% high, 15% medium and 80% low priority. Such a distribution allows the operators to be relaxed for most of the alarms. If one of the (rare) high-priority alarms occurs, the operators could devote their full attention to the alarm. Of course this assumes a proper priority assignment. If a high number of alarms classified as low priority are actually high priority, the priority assignment would not help the operators, even if the recommended priority distribution has been reached.

4.1 Alarms and Events

Table 4–1 Example of a Priority Mapping

OPC severity	OPC severity	Legacy priority	Remapped priority
800–1000	High	1	900
600–800	medium	2	700
< 600	Low	3	500

A proper priority assignment requires lots of often multi-disciplinary know-how. It is not easy and will be very time consuming, as many alarms will require lengthy discussions. But without properly engineered priority assignments, the decision as to which alarms are urgent and important is left to the operators, who in many cases do not have enough experience and background knowledge to make such a decision. It is a cornerstone of the ARC CPAS vision that operators get sufficient support to make consistent decisions, and providing them with well-engineered alarm priority assignments is a good example of how this can be achieved.

Figure 4–5 shows an example of how priority values can be systematically assigned to alarms. The weighted consequence includes an estimated cost for safety, environmental impact, and equipment damage. Depending on the process, other assignments might make more sense. The key point is that the assignment is done systematically from an operator's point of view—not ad hoc—and that the resulting priority distribution is calculated.

Vendor-specific Attributes

In the past, alarms had only a few fixed attributes such as time stamp, tag name and message. The amount of information was limited by the typical output devices, which included 80 character screens or 120 character line printers. The information was enough for a typical post mortem analysis: "What happened when, and in what sequence?"

The OPC AE specification defines several standard attributes but allows implementers of OPC AE servers to define so-called "vendor-specific attributes" where additional information can be attached to alarms. For example, an attribute "AREAS" could contain strings with the names of all areas an alarm belongs to. Vendor-specific attributes can be used as a basis for advanced filters. The content of the vendor-specific attributes can be stored in databases and allows more advanced analysis of past alarms.

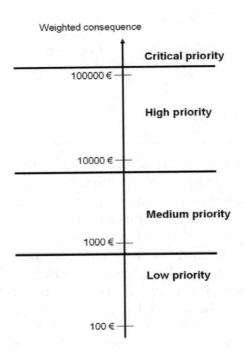

Figure 4-5 Example for a Systematic Priority Assignment as Suggested in EEMUA 191

Categories

Categories define groupings of alarms and events supported by an OPC AE server. The OPC Foundation (OPC Foundation, 2002) suggests using the following categories: Level, Deviation, Discrete, Statistical, System Failure, Device Failure, Batch Status, System Message, Operator Process Change, System Configuration and Advanced Control. But it is up to each OPC AE server to define its own categories. Categories are used to group and filter alarms and events. They are the most common tool to separate alarms for the different user groups, e.g., to route alarms to operators and maintenance staff.

References

Electric Power Research Institute *Advanced Control Room Alarm System: Requirements and Implementation Guidance*; EPRI report 1010076. Palo Alto: 2005.

OPC Foundation *Alarms and Events Custom Interface Standard Version 1.10* (2002). www.opcfoundation.org

Woods, D.D. "The Alarm Problem and Directed Attention in Dynamic Fault Management." *Ergonomics* 38(11), 1995: pp. 2371–2393.

4.2 Common Process Control Application Structure

David Huffman

Introduction

ISA-88, which was originally crafted for batch processes, provides a consistent vocabulary, design process and representations for Process Control and Logical Control. The standard also provides a well thought out process for managing production targets, process control, and procedural control and reporting. With the recent addition of Part 5 to the standard, emphasizing the relationships of Control Modules, Equipment Modules, and Units as well as the coordination control between these elements, ISA-88 now provides a firm foundation to extend the application of these control components to any process.

State Based Control (SBC) is a plant automation control design based on the principle that all process facilities operate in recognized, definable Process States representing a variety of normal and abnormal conditions of the process. Implementing SBC is significantly enhanced and simplified by utilizing the ISA-88 modular element concepts as the basis for organizing the automation applications. State Based Control, implemented with the latest developments in object-based technologies, delivers direct benefits to its adopters in a variety of Operational Excellence categories, benefits that result in productivity increases, higher asset utilization of both people and process, automated responses, and recovery from abnormal conditions. In addition, SBC provides an environment for knowledge capture directly into the control design.

Process States

Process States are characterized by definable differences in processing condition where changes in rates or product grades, or abnormal conditions, dictate

4.2 Common Process Control Application Structure

changes to the automation and control parameters of the process that include, but are not limited to:

- Enabled and disabled alarms. Alarm management becomes part of the design. An alarm is only active when required for the state of the process, which will significantly reduce the nuisance. Current alarm management tools typically mitigate the alarm flow while state-based alarming addresses the root causes of alarm showers.

- Varying alarm limit values

- Loop set-point values, loop configurations (single or cascade), loop tuning for turndown

- Active and inactive interlocks. Like alarming, interlocking generally changes requirements based on the process conditions and those changing requirements become part of the design. Most control configurations are built for a single, specific condition set and then require high level override security features or even on-the-fly configuration changes to deal with any other conditions.

- Providing operator direct access or lock-out for loop operation or tunable parameters. Again, it is not uncommon to want to lock out an operator from manually interfacing with devices in certain situations and then to allow that same operator manual access to some or all device features under other conditions. Traditional designs normally only provide for one or the other. SBC creates an opportunity to include any combination of these requirements into the design of the automation.

- Various running or out-of-service states of individual control modules, equipment modules or units. Inclusion of Units is key beyond many traditional implementations. SBC provides a simpler way to manage control modules and equipment modules than is traditionally available, but such controls have been executed in traditional systems. SBC, through many of the same mechanisms, also provides a convenient method for creating intra-unit coordination control.

Overall production improvement is created by simple situational optimization. By managing the parameters across Process States, optimized process conditions can be maintained during process conditions considered both normal and abnormal. In traditional designs, changes to these parameters are limited or completely ignored, limiting process optimization to only a single, arbitrary normal condition set.

State Based Control

State Based Control is a control design philosophy that embraces a concept where all manufacturing processes operate in a series of definable "states." Generally, SBC is recognized in discrete manufacturing or in batch processing with staged unit operations such as Filling, Heating, Reacting, etc. However, SBC can be applied more generally to all processes, including those normally considered continuous. For many processes, the states may be more generalized as Out-of-Service, Starting, Running, Reduced-Capacity, Shutting Down, etc., but regardless of the labels applied to the states, SBC is a means of providing a far improved solution over traditional, monolithic automation designs.

Key items for State Based Control include:

- Control Module: Generally a single or small grouping of individual control devices, e.g.:
    - Single Valve
    - Pump with suction and discharge control valves
- Equipment Module: Groupings of Control Modules that perform a coordinated control function, e.g.:
    - Can be a single Control Module like a motor that performs agitation.
    - Combinations of pumps, discrete valves, transmitters, and PID control modules to execute a function like "Heating" or "Pressure Control"
- Unit: A functional grouping of Equipment Modules and possibly some individual Control Modules, e.g.:
    - Reactors
    - Distillation Columns

Operational Proficiency and Effectiveness

The Case for Equipment Modules

By the nature of SBC design, the number of items the operator needs to directly control is significantly reduced. At the lowest levels of SBC design, the operator interface is intended to be the Equipment Modules (EMs) and not individual control devices as in traditional designs. The EMs, being composed of one or more Control Modules, are managed by a set of Equipment Module Modes of Operation.

Consider a simple Equipment Module consisting of two motor controls for two pumps (A & B) and two discrete isolation valves for each pump (A_IN, A_OUT, B_IN, B_OUT) for a total of six Control Modules. This Equipment Module might have seven defined Modes of Operation:

1. A Running – B Standby
2. A Running – B Out of Service
3. B Running – A Standby
4. B Running – A Out of Service
5. A & B Out of Service
6. A & B Standby
7. A & B Running

In a traditional design, the operator must open, operate, and close six faceplates to deal with the six devices. *In an SBC design, there is only one faceplate, the Equipment Module faceplate.* When dealing directly with the Equipment Module, the operator need only select the Mode of Operation required, and all control actions are managed by the Equipment Module to deliver the requested state. To perform a simple task such as changing from condition 1 to condition 3 above in the traditional design requires the operator to perform as many as 24 actions between the six devices (open faceplates, select the new Mode of Operation values, execute the new values, close faceplates). The same change using the Equipment Module requires no more than 4 actions (open the faceplate, select the new Mode of Operation value, execute the selection, close the faceplate).

In real-world applications, the amount of reduction in items will vary greatly, but reduction factors of three to six can be considered typical. With devices under the control of a single operator now reaching numbers well in excess of 500 in some cases, reducing the number of items and actions by three to six times is significant progress in addressing operator overload.

Further reductions in items to control can be achieved with extending the State concepts to entire operational units or Unit Modules. By grouping several Equipment Modules and Unit Level Control Modules into a Unit Module, the entire unit can be managed with the application of Process States.

Consider a continuous distillation tower with a single feed, top and bottom product draws, and a two point side draw product option. The Distillation Unit groups together six key Equipment Modules. Based on the concepts discussed previously about Equipment Modules, controlling this distillation tower using that design has already improved the operator's effectiveness as now only six things need to be monitored and controlled instead of a number in excess of 20.

The benefits of Unit Module design appear when true Process State changes are required. Process States for this tower may be wide-ranging but will certainly include categories like Out-of-Service, Starting Up, Full Reflux, Normal Operation with High Sidedraw, Normal Operation with Low Sidedraw, Normal Operation with No Sidedraw, and Shutting Down. Other Process States that could be included are Standby (different than Out-of-Service) and Reduced Operation (caused by downstream unit problems), and could include states for producing different products that are as simple as implementing key set-point changes only.

A simple example of providing for operator effectiveness is changing from any "normal" Process State to Total Reflux due to operational problems somewhere else in the manufacturing train. This change requires manipulation of all of the Equipment Modules in the Unit:

- Feed needs to be stopped.

- Products need to be blocked in and pumps shut down.

- Reboil and Reflux Equipment Modules need adjustment to maintain the Full Reflux status while minimizing energy consumption and possibly reducing or eliminating flaring.

In a traditional implementation, this would require the operator to interface to nearly every loop and valve on the unit; and not just one-time attention items, but a variety of things that need to occur in a proper sequence. This effort is time consuming and diverts the operator's attention away from determining and addressing the root cause of the disturbance elsewhere in the train that required these actions in the first place. This is a significant loss of focus and loss of effectiveness for the operator.

Perhaps more compelling is the lost production due to the time lag in dealing with the root cause of the disturbance, and additional time in getting the train back to full production after the root cause has been addressed. In many plants, a situation of this type often requires other operators to step in and assist in performing all of the necessary actions to focus on both the root cause issue and all of these peripheral items that result, further reducing the effectiveness of the operating personnel.

With the Equipment Module design, things are much improved as there are fewer items to deal with, but a series of things still need to be done that require the operator's attention.

With the Unit Module design of SBC, the operator can access the unit faceplate and make a selection to change Process States from the current conditions to Total Reflux, and all of the necessary control actions will automatically occur in the proper sequence. The operator is free to deal with other critical items and is provided with an environment that maximizes his/her overall effectiveness and the effectiveness of others involved in addressing the root cause of the problem, to correct it, and return the facility to full operation faster. Less time spent at the abnormal condition means more time at profitable, full production. Additionally, during the problem resolution, the process was operating at optimal conditions for the situation, providing the best available profitability for that situation.

Operational Consistency

Operational consistency derives from the constant application of best practices related to implementing key operating procedures. Many facilities try to utilize a variety of methods to reach this goal that range from peer mentoring to detailed procedures. But few of these methods ever really translate into sustainable results as mentors frequently have different methods and skill levels and procedure documents are not always available where and when you need them. So for many facilities, it remains difficult to have every operator execute critical procedures in exactly the same way. In many manufacturing facilities, deviating from a best practices procedure can cause injuries, reduced quality, loss of life, loss of assets, loss of production, and/or create an undesirable environmental impact.

SBC creates a very structured mechanism to allow for critical procedures to be executed the same way, every time they are necessary, without the variability of the human factor. This can be very important in situations where a critical procedure is not frequently executed. One simple example might be lighting-off a furnace or fired heater. In many large, continuous processes that run for years at a time, it is not unusual to find that following a shutdown, many, and perhaps all of the operators now working a unit have never performed such a procedure. Startups are known to be one period of a plant's operating cycle when most of the costly and/or catastrophic mistakes are made. Automating the process provides for consistency and safety in executing the procedure.

Using SBC, and committing to a policy to have the programmed actions updated over time to represent the best practices then known about the operation (just like updating written documents) retains the knowledge of expert operators and provides a means for each new operator to execute their actions just like the experts, following a documented operating procedure. Rather than being faced with the loss of experience that usually leads to lower profitability, plants can make the knowledge available to all who need it, and it can be applied to maintain the same level of financial performance.

Another example of a requirement for consistently providing for a well-defined operating discipline can be found in processes that operate over a wide range of conditions (Process States) that, because of throughput or extreme variations in other operating parameters, need to have PID (proportional-integral-derivative) tuning constants, enabled alarms, alarm limits, interlock conditions, and many other types of control configuration or tunable information changed to create a safe and profitable operating environment. In many cases today, these types of changes are entered manually from checklists or other documents. This creates the opportunity for something to be neglected or entered improperly. Such mistakes can lead to less than optimal performance or to an unsafe condition.

Attempts have been made to use tools outside of the control environment, like spreadsheets, to provide some automated way of making the changes. Maintaining the accuracy of such tools or their ability to work against changing versions of the control system has proven to require the continuous application of engineering resources and contributes a significant addition to the life-cycle ownership cost of the control system. In many facilities that should perform

4.2 Common Process Control Application Structure

these changes, a single set of average "tuning" values is programmed for all conditions and provides less than optimal performance across the range of Process States.

SBC provides a significantly more reliable and profitable way of executing these changes, using the state processing engines at the heart of the individual Equipment Modules and Units. The changes are configured directly into these building blocks, created from core functionality provided by the system supplier, so life-cycle support is designed in. Additionally, in the case of manual entry techniques, if the number of changes is extensive, the time to execute the changes must be considered. SBC can execute the updates instantly if needed, or staggered to address the transitional flow of materials through the process. This capability is directly tied to maintaining prime product quality and minimizing off-specification or lower quality materials during these transitional periods.

Engineering Efficiency

SBC provides an automation design philosophy that is applicable to nearly all automation situations: continuous, batch, discrete, or any combination of processing requirements. CPASs increase the efficiency of engineering SBC applications and reduce life-cycle support efforts with:

- Modular design that creates opportunities for significant re-use.

- Utilization of standard state engine features, eliminating custom logic code creation and maintenance.

- Standardized implementation and maintenance techniques across all process types, which eliminates the need for segregated pools of specialists to execute projects or maintain systems throughout the application's life cycle.

- Techniques that are available to reduce specification and documentation impacts.

Plants that process materials in sequential or batch processes have benefited for many years by implementing software features like Control Modules, Equipment Modules and Units. Applying these same equipment elements with SBC to continuous control applications offers this same basic set of savings.

The degree of re-use for most companies can be widespread at the Control Module level, somewhat less at the Equipment Module level, and may be limited to certain key processes at the Unit level. But where common sets of Control Module and Equipment Module components can be applied generically to both continuous and batch processing, there will be a significant reduction in upfront engineering hours.

Modular design is best implemented with a system that provides true object-based functionality. Benefits can be derived from template-based implementations, but the life-cycle support requirement when templates are used will be significantly higher than with object-based systems. Objects provide a true design, test, and document-once environment, not only at the original design point, but also when updates and changes are required, significantly reducing validation and testing throughout a life cycle of inevitable changes. Additionally, object functionality, by design, supports the parent-child relationships exhibited with using Units, Equipment Modules, and Control Modules design, eliminating any special associative design requirements for making the associations for drill-down functionality within the operational environment that are often typical of template-based designs.

Reduce Custom Coding with Standard Features

In addition to the reuse benefit in continuous applications, state based coordination design, using the sequencing and state management features, is significantly more cost effective than trying to perform the same functionality with traditional coding methods.

Most implementations of the types of control identified in the earlier examples were previously done with a significant amount of if-then-else or complex ladder logic coding, resulting in applications that were very costly to implement initially, and generally increased in complexity (and in the costs to maintain them) over time. In the case where a plant has remained unchanged for many years, and a process expansion or major process retrofit is required, there are often no resources remaining that readily understand the original coding. So not only is there a need for engineering hours to modify the old code for the new process updates, but a significant number of hours must be added to first interpret, understand, and document what already exists.

SBC, designed around the use of a sequencing engine configured to manage Process States, significantly reduces or eliminates the need for custom logic coding structures (Figure 4–6). The sequence engine design for Process

4.2   Common Process Control Application Structure   233

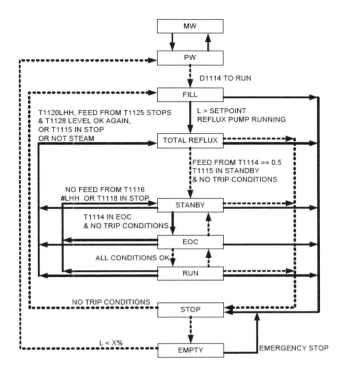

Figure 4–6   Example of State Based Sequence Control Design

State management in Units and for Modes of Operation in Equipment Modules provides an easy-to-configure, and easy-to-visualize, means of programming all of the state-managed parameter manipulation that can be required by a highly complex process, significantly reducing the time associated with standard coding processes. The use of the sequence engine also provides a form of self-documenting information that can be easily tested and verified at initial project testing steps (including Factory Acceptance Test and Site Acceptance Test), and provides a simplified base for understanding the code for maintenance, updating, or expansion requirements.

The operator plays a key role here also. The sequence view of process control is one that is easy for an operator to understand.

It is important that the operator can easily access the sequence and troubleshoot it directly from their workplace with appropriate visualization tools (Figure 4–7). Having operators read structured text or ladder logic for understanding and troubleshooting is difficult, and systems that require operators to

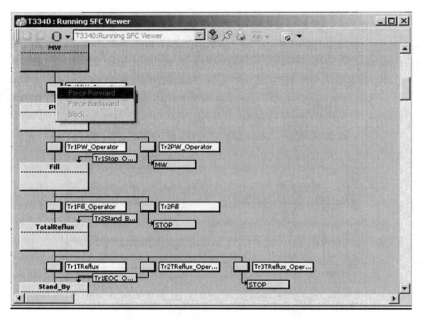

Figure 4–7   SFC Viewer Example: Unit Process States

use engineering tools for this purpose greatly increase the inefficiencies, and increase the chance of errors or misinterpretation.

Specification and Documentation Improvements

There are two actions related to a project life cycle that are frequently taken for granted or often not performed to expectations or desired completeness: Functional Specification and Life Cycle Documentation. To be completed properly, both can consume large quantities of engineering hours. They often do not get completed properly because those hours are not available and take second or lower priority to other daily responsibilities for the resources that can most effectively perform these functions.

Whether the team that will do the automation system configuration is the system supplier, a system integrator, or an internal group of end user engineers and technicians, that group is normally supplied reference documentation:

- P&IDs and/or PFDs (Piping and Instrumentation Diagrams; Process Flow Diagrams)

4.2 Common Process Control Application Structure

- Instrument list
- Alarm lists
- Interlock tables
- A "prose" style document normally titled Functional Specification

Functional Specification documentation is normally a descriptive dissertation about what one person thinks the process is supposed to do and how it is to be controlled. When it is time to develop the control code, a different person interprets that description in conjunction with the other items listed above and creates a control configuration. Often a Design Specification is written as a means of showing the interpretation of the Functional Specification. Additionally, a Testing Specification may also be required to provide detail on what and how testing will be done to show that the design meets the functional requirements. Writing, interpreting, and reviewing documents is time consuming and costly.

At a frequency much higher than is commonly understood, problems in the translation from specification to configuration are not found until actual factory or site acceptance testing. When projects really go bad at this point, time is lost that can lead to a late delivery and a corresponding overall project startup delay, engineering hours are required to correct the problems, and project managers can spend countless hours arguing over fault and compensation.

SBC provides for standardizing on specification information. If the end user is committed to using SBC, then the documentation they provide can be structured in a way that all suppliers or integrators can understand and respond with little or no confusion. There will likely be three basic types of documents in addition to P&IDs/PFDs:

- Listing of Control Module types required for the project, with description of functionality

- Equipment Module configuration table (Figure 4–8) for each module, providing Mode of Operation state condition details for each Control Module contained within the module

Unit configuration table (Figure 4–9) identifying available Unit Process States and appropriate Mode of Operation for each Equipment Module in that Process State.

Equipment Module Template						
Unit Association:	Distillation					
Equipment Module Name:	Reflux EM					
Device Type:	MOTOR_1_SP	PID	PID	CTRL_VLV	Step Used	
Device Description:	Duty/Stby	Reflux Ctrller	Level Ctrller	Reflux CV		
Device Tag:	P-1113	D1114_Rflx_PID	D1114_Lev_PID	T1114_Rflx_FV		
MOP						
MOP=1 Off	Off	Track = On	Track = On	Closed	M.W.	Yes
	ILock = None	(Value = 0)	(Value = 0)	ILock = None	P.W.	Yes
	Permis = Lock	AlmHH = Inactive	AlmHH = Inactive	Permis = Lock	FILL	
		AlmH = Inactive	AlmH = Inactive	Op_Lock = Yes	TOTAL REFLUX	
		AlmL = Inactive	AlmL = Inactive		STAND-BY	
		AlmLL = Inactive	AlmLL = Inactive		EOC	
					RUN	
					STOP	Yes
					EMPTY	
MOP=2 Recirc	On	Track = On	Track = On	Closed	M.W.	
	ILock from LT	(Value = 0)	(Value = 0)	ILock = None	P.W.	
	Permis = None	AlmHH = Inactive	AlmHH = Inactive	Permis = Lock	FILL	Yes
		AlmH = Inactive	AlmH = Inactive	Op_Lock = Yes	TOTAL REFLUX	
		AlmL = Inactive	AlmL = Inactive		STAND-BY	
		AlmLL = Inactive	AlmLL = Inactive		EOC	
					RUN	
					STOP	
					EMPTY	
MOP=3 Total Reflux	On	Track = Off	Track = Off	Tracking	M.W.	
	Ilock - None	Master (Level) - Slave	Master to Reflux Ctrller	ILock = None	P.W.	
	Permis = None	AlmHH = Inactive	AlmHH = Inactive	Permis = None	FILL	Yes
		AlmH = 80%	AlmH = 80%	Op_Lock = No	TOTAL REFLUX	Yes
		AlmL = 20%	AlmL = 20%		STAND-BY	Yes
		AlmLL = Inactive	AlmLL = Inactive		EOC	
					RUN	
					STOP	
					EMPTY	
MOP=4 Take-Off	On	Track = Off	Track = Off	Tracking	M.W.	
	ILock from LT	Single Loop	Master to Reflux Ctrller	ILock = None	P.W.	
	Permis = None	PV from FT	SP from OP	Permis = None	FILL	
		SP via MMS from T-1115	PV from LT	Op_Lock = No	TOTAL REFLUX	
		AlmHH = Inactive	AlmHH = 90%		STAND-BY	
		AlmH = 85%	AlmH = 75%		EOC	Yes
		AlmL = Inactive	AlmL = 10%		RUN	Yes
		AlmLL = Inactive	AlmLL = 5%		STOP	
					EMPTY	
MOP=5 Empty	On	Track = On	Track = On	Closed	M.W.	
	ILock from LT	(Value = 0)	(Value = 0)	ILock = None	P.W.	
	Permis = None	AlmHH = Inactive	AlmHH = Inactive	Permis = Lock	FILL	
		AlmH = 80%	AlmH = Inactive	Op_Lock = Yes	TOTAL REFLUX	
		AlmL = 20%	AlmL = 10%		STAND-BY	
		AlmLL = Inactive	AlmLL = 5%		EOC	
					RUN	
					STOP	
					EMPTY	Yes

Figure 4–8 Example Equipment Module Configuration

Providing information in these or similar formats aids in reducing engineering hours and increasing the effectiveness of anyone that needs to use this information.

4.2 Common Process Control Application Structure

UNIT PROCESS STATE EQUIPMENT MODULE SETUP TEMPLATE				
UNIT	Distillation			
Equipment Module	Reflux EM	Reboiler EM	Bottom EM	Transfer EM
EM Description	E-1110	E-1111	Bottom Level	Sidedraw
Step				
M.W.	Off	Off	Off	PCC_OFF
P.W.	Off	Pressure	Off	PCC_OFF
Fill	Off Level > SP1 = RECIRC Level > SP2 = Total Reflux	Fixed SP > 0 & LI-1112 > LowLevel = Flow	Recirc LI-1112 > SP = Level	PCC_OFF
Total Reflux	Total Reflux	Flow	Recirc	PCC_OFF
Stand-by	Total Reflux	Flow	Level	PCC_OFF
EOC	Take Off	Pressure	Level	Line1 or Line2
Run	Take Off	Ratio	Level	PCC_OFF
Stop	Off	Pressure	Off	PCC_OFF
Empty	Empty	Pressure	Empty	PCC_OFF

Figure 4–9   Example Unit Configuration Table

Bibliography

ISA-88.01 *Batch Control Part 1: Models and terminology*. Research Triangle Park: ISA, 1995 (R2006). See also IEC 61512.

ISA-88.00.05 *Batch Control Part 5: Implementation models and terminology for modular equipment control* (working draft standard). Research Triangle Park: ISA.

Brandl, D. "Design Patterns for Flexible Manufacturing." Research Triangle Park: ISA, 2006.

Chappell, D.A. "Using S88 Batch Techniques to Manage and Control Continuous Processes." Woodcliff Lake: World Batch Forum, 2003.

4.3 Remote Operation and Service

Martin Hollender

Introduction

Due to globalization, many companies operate production facilities spread all over the world. This often means that the required process and automation expertise to solve non-routine maintenance and troubleshooting problems is not available locally at the plant. In fact, some problems can best be solved by involving several experts who might be located in different parts of the world. If a problem can be solved remotely, this has important advantages:

- Reduced downtime when problems are fixed faster. Remote service leads to faster response times because:
    o Travel time can be avoided.
    o It can provide follow-the-sun, around-the-clock service by using world-wide service centers in different time zones.
    o It can quickly bring in the required level of expertise on a case-by-case basis.
- Reduced travel cost with lower emissions
- More economical service from countries with lower labor cost
- The possibility of bringing in higher level expertise from different places. Many CPAS suppliers have created centers of excellence focusing on different specialties, which together can offer very competent service.
- Remote service enables more proactive maintenance. Remote experts can identify developing problems at a very early stage and have them fixed before they develop into real problems.

4.3 Remote Operation and Service

The key issues of remote access are security (see Section 2.7) and access management. Remote service is a standard practice in the IT (Information Technology) industry because many companies have outsourced the administration of their IT infrastructure to external specialists—and the IT infrastructure is an important basis for CPAS. Some services such as security patch management (see Section 2.7) require fast response times, because otherwise the system is left vulnerable. Another possible service is that the CPAS vendor automatically installs bug fixes so that already-solved problems from other plants cannot re-occur.

Diagnostic messages—for example, that the hard disk of a CPAS server is getting full or that a motor is getting too hot—can be issued as email or SMS (Short Message Service) to remote technicians.

A second important trend is to operate plants remotely because direct operation is too expensive, either because the environment is dangerous and harsh, like at an offshore oil platform, or because the process is running at a high level of automation, requiring very little human control and supervision.

Telecommunication and Computing Infrastructure

The worldwide telecommunication and computing infrastructure has been dramatically improved during the last decades. Broadband Internet access and satellite connections are commonplace today. Cheap, fast, and secure remote connections are now a commodity. Teleconferences with participants from several countries are now daily practice in many companies. The Internet has dramatically changed interaction between people from distant places. The IT infrastructures of many companies are already managed by external service providers, with most of the interactions done remotely. This can serve as a model for the service of a CPAS.

Standard operating systems have features like remote desktop access, telnet, or remote shell, which allow participants to work from distant computers. Commercial remote access and collaboration services are offered by companies including Microsoft, Citrix, Cisco and Adobe.

Early remote access technologies such as point-to-point connections using modems had the disadvantage that they cannot be centrally managed, so that it was often not clear which and how many modem connections existed in a plant. It was not uncommon that service technicians changed companies and took modem numbers and passwords (which were often exchanged between colleagues) with them.

Today CPAS vendors offer centralized access management solutions. They define who is authorized to access a remote plant, and audit each access at one central place. Usually these solutions are web-based. As HTTP (hypertext transfer protocol) connections cannot be established from outside a firewall, the solution includes software agents running inside the plant which register in the remote access management server over the Internet. These agents can also be used to monitor the CPAS and if necessary send an alert to the remote service center. Once a remote technician needs to access the plant behind the firewall, the remote access server can use the registered outgoing HTTP request to establish an ingoing connection.

Remote Service

Remote service is offered by several CPAS vendors. For example, services for security patch management (see Section 2.7), loop tuning (see Section 5.2), and fault diagnosis and alarm management (see Section 6.2) can be provided remotely. Data is collected in the plant and then sent to a remote service center for further analysis. The results are sent back as an email or can be accessed in a web portal.

Even when travel cannot be avoided, for example because devices need to be physically inspected, remote access can help technicians diagnose the situation in advance and prepare the necessary tools and spare parts.

In some cases the remote service center has permanent access to devices or even a whole plant, whereas in other cases plant personnel open remote access for the external service experts on a case by case basis.

Intelligent Device Management (IDM) software allows equipment manufacturers to proactively monitor and manage remote devices and systems (see also Section 5.4 on plant asset management). Specialists from the CPAS vendor can use IDM to track equipment usage and performance, push software upgrades, perform maintenance checks, and execute repairs. IDM technology allows the vendor to offer more support at less cost and therefore maximize the return on assets. With the help of IDM it becomes possible to monitor a whole population of devices under real world conditions. Device manufacturers can use this feedback data to improve their device designs. The data can also be used for the optimization of the maintenance strategy.

Remote Operation

For dangerous locations like offshore oil platforms, remote operation allows the operators to supervise the process from an onshore location (Figure 4–10). This is not only much safer but also cheaper, as every offshore man-day is very expensive. Working onshore allows more regular working times, and manning levels can be optimized. While it is usually not possible to run a platform totally unmanned, the number of offshore workers can be significantly reduced. Similar benefits exist for remote operation in other dangerous or hostile locations.

It is evident that security (see Section 2.7) is an essential precondition for remote operation and service. Nobody wants plants to be remotely controlled by terrorists. Today there are already a significant number of remote operation centers in place, which has proven that connections are fast, reliable and secure. These operation centers are well equipped not only for remote service, but also for the collaborative solution of operational problems. High-level experts who might be working in different parts of the world can be involved to solve difficult questions.

Centralized Control and Supervision

As an additional use case for remote operation, some highly automated processes have relatively little need for human control and supervision. In such cases several processes can be run from one centralized control room. One 24h/day staffed centralized control room can supervise several, and sometimes even rather unrelated, processes. Multisystem integration enables seamless global access to real-time values, alarms & events, and historical data across whole constellations of systems. Without a CPAS capable of integrating multiple systems, the integration would require costly project specific programming, which often results in error-prone and low-performing functionality. Data transfer between systems is optimized. The network between subscribers (supervising systems) and providers (supervised systems) can be anything from a high-speed LAN (Local Area Network) at 1 GBit/s down to modem connections with a speed of 128 kBit/s.

A related trend is the integration of formerly separate process and electrical automation in power distribution and substation systems, which yields significant savings in terms of costs and manpower. The CPAS needs to support IEC 61850, the global communication standard for Power Distribution and

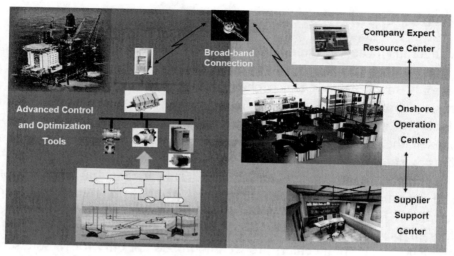

Figure 4–10 Remote Control of an Offshore Oil Platform

Substation Automation. Operators can view and control both systems from one user interface. Integrated process and electrical systems make it possible to run power management applications in the process automation system, providing a robust, reliable power control system that reduces the possibility of power interruptions or blackouts while optimizing all operations and assets, including the ones of the electrical part of the plant.

Bibliography

Biehl, M. Prater, E., and McIntyre J. "Remote Repair, Diagnostics, and Maintenance." *Communications of the ACM*, Volume 47, Issue 11, 2004.

Cheever, G. and Schroeder, J. "Remote Service." *ABB Review Special Report Automation Systems*, 2007. Available at http://bit.ly/Lybsh (retrieved 01-May-09).

CHAPTER 5

5.1 Advanced Process Control

Alf J. Isaksson

Introduction

Let us start by defining what we mean by Advanced Process Control (APC) in this context:

> *APC is all control and on-line optimization in a process industry that operates in closed-loop, using automatic feedback from process measurements, and is more advanced than single PID controllers.*

This means that APC includes all types of multivariable controllers such as model predictive control (MPC), internal model control (IMC), and linear quadratic gaussian (LQG). However, it also means that controller structures such as feedforward, ratio, decoupling, and midrange control are also regarded as APC.

Planning and scheduling optimizations that are part of a manufacturing execution system (MES) should not be seen as part of APC.

What is meant by "process industry" was defined already in the first chapter. It suffices here to point out that the common denominator is that the controlled variables are typically continuous in range rather than discrete, which is the case in the discrete manufacturing industry. For this reason, power plant control is typically also regarded as process control despite the fact that no products, in a conventional sense, are produced. Discrete variables do, however, exist also in process industries and may also be controlled as will be discussed in some examples below.

Although, as stated above, APC covers a broad range of control approaches, the model-based methods get more and more attention. In fact, in a 2002 report, the ARC Advisory Group (O'Brien, 2002) referred to this decade as the model-centric decade. In some industries model-based control has been used for more than 30 years. For some other industries, for example, pulp and paper, power

5.1 Advanced Process Control

generation, cement, etc. the ARC statement certainly applies, since in recent years we have seen more and more applications of model-based solutions.

Therefore, in this chapter we will focus on the main model-based method deployed today, viz. model predictive control (MPC). Unlike most other control techniques, MPC was first developed by industry in the 1970s, and not as academic research. See Cutler and Ramaker (1980) and Richalet et al. (1978) for a couple of early papers. For a more recent textbook, see Maciejowski (2002).

Today, when linear MPC is finally making its way into the process industries on a wider scale, we are already seeing the first applications of nonlinear MPC techniques, as well as MPC applications with so-called hybrid models. There is also a general trend of moving up the value chain, focusing on plant-wide production optimization, as well as enterprise-wide planning and scheduling; hence making the borders in the traditional automation hierarchy less and less distinct.

Technical Background

As the name model predictive control indicates, a crucial element of an MPC application is the model on which the control is based. Therefore, before a controller can be implemented a model has to be established. There are two main alternatives available for obtaining the model:

- Deriving a model from first principles using the laws of physics, chemistry, etc., called white-box modeling

- Estimating an empirical model from experimental data, called black-box modeling

In general, a white-box model becomes a set of nonlinear Differential and Algebraic Equations (DAE)

$$\dot{x}(t) = f(x(t), u(t))$$

$$y(t) = h(x(t), u(t))$$

where $y(t)$ denotes the measured process variables, in a CPAS typically referred to as PV (process variables) or CV (controlled variables). The manipulated variable, i.e., the output of the MPC, is denoted $u(t)$. In a CPAS, this would be called MV (manipulated variable) or CO (control output), but in most cases this corresponds to set points for a level of PID controllers between the MPC and the actuators of the actual process, and as such is instead called SP (set point). Finally, the internal variable $x(t)$ is what is usually referred to as the state of the system. This type of model is also called a (nonlinear) state-space model.

A black-box model, on the other hand, is typically linear, but most often also discrete in time:

$$x_{k+1} = Ax_k + Bu_k$$

$$y_k = Cx_k$$

Here the integer k denotes the kth time index for which the signal value is available, i.e., at time $t=kT_s$, where T_s is the sampling interval. Hence, we have, for example,

$$x_k = x(kT_s)$$

However it is modeled, the core of MPC is optimization. In each iteration of the control, i.e., any time a new measurement is collected, two optimization problems have to be solved (both using the model as an equality constraint): one using past data to estimate the current state vector x_k and one to optimize the future control variables. When solving the forward optimization problem, a number of future values of the manipulated variables are calculated. However, only the values at the first time instant are transmitted to the underlying process. At the next time instant the optimizations are repeated, with the optimization windows shifted one time step. This is known as receding horizon control (see Figure 5–1), and is in fact what makes this a feedback control method. Performing optimization just once would correspond to open-loop control.

For the state estimation, the optimization target is to obtain the best estimate of the internal variable x using knowledge of y and u, to be used as a starting point for the forward optimization. This can be done using a Kalman filter (for an old classic see Anderson and Moore, 1979)—or if the model is nonlinear, an extended Kalman filter—where a stochastic modeling of process and measurement noise is applied. A Kalman filter is a recursive method, meaning that it takes only the most recent values of y_k and u_k to update the previous esti-

5.1 Advanced Process Control   247

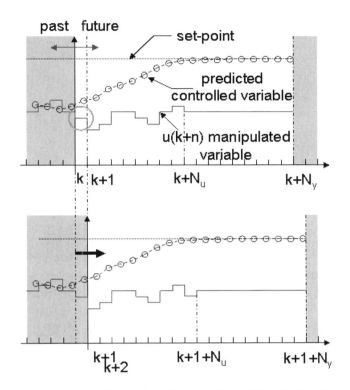

Figure 5–1 Illustration of Receding Horizon Model Predictive Control

mate $\hat{x}_{k-1}$ to the new one $\hat{x}_k$; hence, it does not actually solve an optimization problem on-line.

With today's access to more computational power, a much newer and increasingly popular approach is to use the so-called moving horizon estimation (Rao, 2000). With this approach, the process and measurement noise are introduced using slack variables w and v in a discretized version of the nonlinear model

$$x_{k+1} = g(x_k, u_k) + w_k$$

$$y_k = h(x_k, u_k) + v_k$$

or in the linear case as

$$y_{k+1} = Ax_k + Bu_k + w_k$$

$$y_k = Cu_k + v_k$$

The moving horizon estimation then corresponds to minimizing V

$$\min_{x_k} V = (x_{k-M} - \hat{x}_{k-M})^T P^{-1}(x_{k-M} - \hat{x}_{k-M}) + \sum_{n=k-M}^{k} w_n^T R_1^{-1} w_n + v_n^T R_2^{-1} v_n$$

subject to constraints as, for example,

$$x_{\min} \leq x_n \leq x_{\max}$$

Here P, R_1 and R_2 are weight matrices, used for the tuning of the estimator, which have a similar interpretation and importance as the estimate and noise covariance matrices in Kalman filtering.

As indicated, the optimization for moving horizon estimation is typically done over a horizon of data $[t-MT_s, t]$, where t is the current measurement time. Since this time interval is in the past, we assume access to historic values of the applied manipulated variables u_k and the measured process variables y_k. The first penalty term in the criterion is called the arrival cost. Its function is to create a link from one optimization window to the next, where $\hat{x}_{k-M}$ denotes the estimate for this particular time instant from the optimization run at the previous cycle.

The state estimation produces a starting point for the optimization of future manipulated variables, where future set points r_k are compared with the controlled process outputs y_k, calculated by use of the mathematical process model. A formulation of the optimization objective may be, for example,

$$\min_{\Delta u_k} J = \sum_{n=0}^{N_y} (r_{k+n} - y_{k+n})^T W_y (r_{k+n} - y_{k+n}) + \sum_{n=k}^{k+N_u} \Delta u_{k+n}^T W_u \Delta u_{k+n}$$

subject to, for example,

$$y_{\min} \leq y_{k+n} \leq y_{\max}$$

$$u_{\min} \leq u_{k+n} \leq u_{\max}$$

5.1 Advanced Process Control

$$\Delta u_{min} \leq \Delta u_{k+n} \leq \Delta u_{max}$$

Here the optimization is done using the increments of the manipulated variable $\Delta u_{k+n} = u_{k+n} - u_{k+n-1}$ as free variables, which introduces integral action in the controller.

Notice that in the optimization problems described above the model (non-linear or linear) should be considered as an equality constraint. We will not go into how these optimization problems are solved. Let us just point out that depending on the objective function and constraints (most importantly the model) different types of optimization problems result, leading to different types of optimization solvers being needed. For example, a quadratic objective together with linear constraints corresponds to a quadratic programming (QP) problem, whereas a nonlinear objective or nonlinear model yields a nonlinear programming (NLP) problem which, of course, is much more difficult to solve. The latter case is usually referred to as a nonlinear MPC problem (NMPC).

Application Examples

To illustrate the benefits of MPC, this section contains four relatively recent industrial examples where different types of MPC controllers have been successfully applied.

Cogeneration of Steam and Power

At Point Comfort, Texas, USA, Alcoa has a large refinery where bauxite is converted to alumina. Since this is a very energy consuming process, Point Comfort has its own powerhouse with multiple boilers, turbines, and steam headers (see Figure 5–2 for a system overview). Most of the electrical energy needed is produced in-house, but electricity is also purchased from the local power grid.

With varying prices of electricity and fuel (natural gas) the first problem is to determine the optimal mix of in-house versus purchased energy. This is now done by solving a mixed integer linear program every 15 minutes using current fuel and electricity prices downloaded from the Internet.

The results from the steady state optimization are fed to an MPC calculation which runs with a much faster cycle (< 10 s). The MPC is based on an

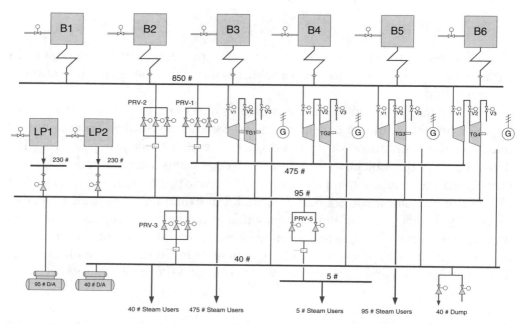

Figure 5–2 The Alcoa Point Comfort Power Plant

empirical linear dynamic state space model, and delivers 28 manipulated base control set points.

The system was commissioned in 2005 and immediately led to greatly improved process stability; for example, an 80 percent reduction of steam pressure standard deviation (see Figure 5–3). A one-percent saving in overall energy cost was verified, giving a six-month payback time. More details on this installation are provided in Valdez et al. (2008).

Power Boiler Start-up

If the process to be controlled has significant nonlinearities, it may not lend itself very well to a purely empirical modeling. An example where first principles models have been used successfully is optimal start-up of fossil fueled steam power plants.

The background is that in the current de-regulated power market, these power plants are not used only for base load, but also to help regulate widely varying loads, and hence encounter many more stops and starts than in the past. The start-up time for a power plant is highly constrained by thermal stresses,

5.1 Advanced Process Control

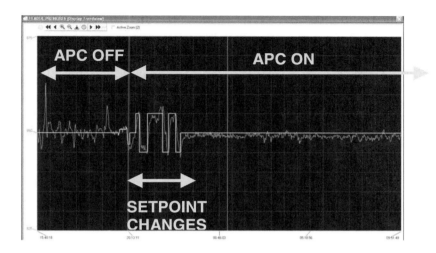

Figure 5–3   Steam Pressure Control, with and without MPC

Figure 5–4   The Weiher III Power Plant, where the First Pilot was Installed

i.e., excessive temperature gradients in thick-walled parts of the boilers may lead to cracks in the material.

Given a model and on-line measurements it is, however, possible to calculate the actual thermal stress. Therefore, an object-oriented boiler model was developed; see Figure 5–5. The model was implemented in the modeling language Modelica (Fritson, 2004), using the software Dymola. The total model

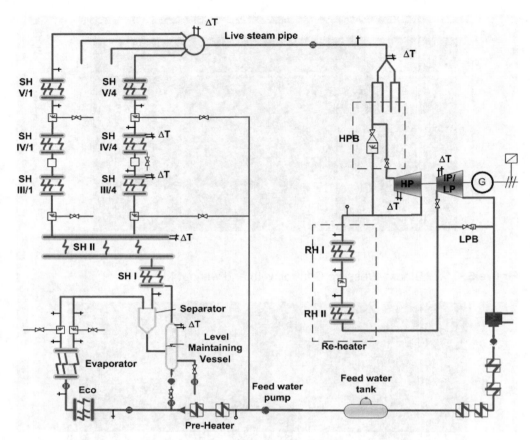

Figure 5–5   Overview of the Object-Oriented Model for Boiler Start-up

comprises 68 components with a total of 1229 differential-algebraic equations (DAE). There are five constrained outputs and two optimized inputs: fuel flow rate and HP (high pressure) bypass valve position.

Dymola converted the model into a set of ordinary differential equations (ODE). This model was then integrated into the distributed control system together with an optimization module. This way the two inputs were optimized to get the fastest possible start-up, subject to not violating the constraint on thermal stress.

Figure 5–6 shows two warm starts (down time one weekend) for a 700 MW coal-fired power plant. With NMPC the start-up time could be reduced by about 20 minutes, and the start-up costs by about 10 percent compared to a well tuned conventional control scheme. For more details, see Franke and Vogelbacher (2006).

5.1 Advanced Process Control

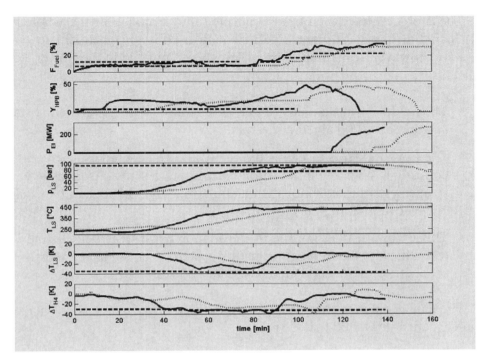

Figure 5-6 Measurements from Start-ups Using Conventional Control (dotted) and NMPC (solid). From top to bottom: Total fuel flow rate, HP bypass valve position, generated electrical power, live steam pressure and temperature, and thermal stress in live steam pipes and HP headers. The dashed lines show constraints

This technology has now been installed at seven power plants, with three more installation projects ongoing. The typical fuel saving for a single start-up is between 10 and 20 percent. With 50-150 start-ups per year, this corresponds to 0.8-8 million kWh per installation.

Pulp Mill Optimization

An NMPC technique was used also to optimize the pulp production and chemical balance at Billerud Gruvön in Sweden. Here a library of models has been developed for the various process objects in the pulp mill, including chemical recovery: pulp digester, pulp tank, white liquor tank, black liquor tank, evaporator, recovery boiler, etc.

These objects have then been connected into a model for the whole pulp mill, including three different digesters with a common chemical recovery. There are, however, some important differences compared to the Power Boiler application:

- The model is substantially larger, having more than 250 dynamic states and another 2000 algebraic variables.

- The DAE model is discretized directly and considered as an equality constraint. With an optimization horizon of 15 steps this leads to a sparse system of equations with approximately 30,000 variables that is solved every 30 minutes.

- This optimizer does not run in closed loop but instead delivers the result back to the mill's information system as decision support for the production engineers (Pettersson et al., 2006).

The annual savings from this system in the reduced purchase of make-up chemicals are at least 500,000 USD.

Hybrid Control

If, in addition to continuous variables, the optimization involves some variables of on/off character—often numerically represented by 0/1—it becomes significantly more complex. Nevertheless, academic researchers have developed formulations for these so-called mixed logic and dynamic (MLD) problems that can be solved by standard solvers for mixed integer programming; for example, see Bemporad and Morari (1999).

An example is the optimal operation of the Iluka Resources Narngulu synthetic rutile plant in Western Australia; see Clark (2008). Using a prediction horizon of 43 hours, the system simultaneously solves the tasks of optimizing the kilns in the dry section as well as the schedules for the start and stop of the aeration batch tanks in the wet section.

The installation resulted in reductions of both material and energy use, giving annual savings of 1 million USD. For the Iluka Resources plant, this corresponded to a project payback time of less than six months.

Clearly, MPC is well established in some industries (refineries, petrochemical), but is also growing rapidly in many other areas (power generation, pulp &

paper, cement, etc.) Hopefully the examples provided have convinced the reader that the benefits make it interesting to apply MPC to his/her industry.

Conclusions

The aim of this chapter was to provide an introduction to advanced process control in general, and in particular the specific approach called model predictive control. A brief technical background of MPC was given, followed by four application examples meant to illustrate the large economical potential of the technology.

What, then, are the costs associated with applying MPC? The main effort for these model-based applications is, not surprisingly, the development and fine tuning of the model. What type of model to go for is largely dictated by the application itself. If

- there are many similar processes where one can deploy almost the same model;
- the process model can be in some form of generic description, so that one can reuse components between projects;
- it is possible to use object-oriented modeling techniques where the model is configured from a library via a graphical user interface using "drag and drop;"
- process domain expertise is likely to be available for maintenance during the life cycle of the model;
- all main physical phenomena are known, even if some parameters have to be estimated;

then physical (white-box) modeling is an attractive alternative. Otherwise, the use of black-box modeling is probably the recommended approach. Sometimes it is attractive to use a combination of the two techniques, which is then called grey-box modeling.

Needless to say, much of the effort by suppliers of MPC products goes into providing software support that improves engineering efficiency. To really make MPC part of the CPAS concept, MPC engineering should have a vision of becoming as simple as basic (PID) automation engineering.

Acknowledgments

The author is grateful to a number of colleagues for valuable comments and for providing background material. In particular I would like to thank Messrs. Rüdiger Franke, Eduardo Gallestey, Sebastian Gaulocher, Ted Matsko, Lars Ledung, and Jens Pettersson.

References

Anderson, B.D.O. and Moore, J.B. (1979). *Optimal Filtering.* Upper Saddle River: Prentice-Hall.

Bemporad, A. and Morari, M. (1999). "Control of Systems Integrating Logic, Dynamics, and Constraints." *Automatica* 35, No. 3, pp. 407–427.

Clark, D. (2008). "Optimum Process Scheduling." *International Mining*, January.

Cutler, C.R. and Ramaker, B.L. (1980). "Dynamic Matrix Control—A Computer Control Algorithm." *Proceedings Joint American Control Conference*, San Francisco.

Franke, R. and Vogelbacher, L. (2006). "Nonlinear Model Predictive Control for Cost Optimal Startup of Steam Power Plants." *Automatisierungstechnik* Vol. 54, No. 12.

Fritzson, P. (2004). *Object-oriented Modeling and Simulation with Modelica 2.1"*, Piscataway: IEEE Press.

Maciejowski, J.M. (2002). *Predictive Control with Constraints.* Upper Saddle River: Prentice-Hall.

O'Brien, L. (2002). *Total Automation Business for the Process Industries World Wide Outook – Market Analysis and Forecast through 2006.* Boston: ARC Advisory Group.

Pettersson, J., Ledung L., and Zhang X. (2006). "Decision Support for Pulp Mill Operations Based on Large-scale On-line Optimization." Preprints of *Control Systems 2006*, Tampere, Finland, 6–8 June.

Rao, C.V. (2000). *Moving Horizon Strategies for the Constrained Monitoring and Control of Nonlinear Discrete-Time Systems*. Ph.D. Thesis, University of Wisconsin.

Richalet J., Rault, A., Testud, J.L., and Papon, J. (1978). "Model Predictive Heuristic Control: Applications to Industrial Processes." *Automatica*, Vol. 14, pp. 413–428.

Valdez, G., Sandberg, D.G., Immonen, P., and Matsko, T. (2008). "Coordinated Control and Optimization of a Complex Industrial Power Plant." *Power Engineering Magazine*, November pp. 124–134.

5.2 Loop Tuning

Alf J. Isaksson

Motivation

Wherever it is implemented, a CPAS is a fairly large investment, and a typical process industry plant may have thousands of control loops. It is a sad fact that too often the investment stops there, without the proper tuning of all these loops. Not only should the control loops be tuned for optimal performance at commissioning, but maintaining good control over the life cycle of a process plant is a never-ending task. A loop that was once well tuned may not be anymore, due to wear and tear or changes in the process, or a shift in operating point, all of which may lead to different dynamics of the process.

The bulk of the level one controllers in process plants are still PID, and for good reasons such as,

- PID controllers often give almost optimal performance for many industrial processes

- Given the large number of loops, the need for only a few parameters to tune

The derivative (D) part of the PID controller is, however, often underutilized, effectively making these controllers PI controllers. This may be because there is an industry fear that derivative action is too noise sensitive. With proper filtering of the D part, the noise can be taken care of and PID in many cases yields significantly improved performance compared to PI control; see Isaksson and Graebe (2002).

Given how old the topic is, PID tuning is still a surprisingly active research area. There were, for example, special sections published by *IEE Proceedings—Control Theory & Applications* in 2002 and by *IEEE Control Systems Magazine* in 2006.

Technical Background

When the PID controller first appeared, tuning was entirely manual, mainly because the tuning knobs had no scales on them. In interviews long after they designed their famous tuning rules, Ziegler and Nichols commented that they always thought their main contribution was that they introduced scales on P, I, and D! Having achieved this, it was a natural corollary to design some tuning rules, but to Ziegler and Nichols this was secondary.

The PID controller comes in many different versions, with different parameterizations. We have no space here to go through all of them, but instead we will stress the importance of reading the manual (especially for the D part) before deploying any controller setting calculated with any of the methods mentioned below.

A very common (perhaps today the most common) parameterization is the so-called ideal form where the controller output $u(t)$ (in a CPAS typically referred to as, e.g., OP, MV, CO or OUT) is calculated as

$$u(t) = K_C \left(e(t) + \frac{1}{T_I} \int_0^t e(\tau) d\tau + T_D \frac{d}{dt} e(t) \right)$$

Here $e(t)$ is the control error calculated as the deviation between a reference value (or set point) $r(t)$ (in a CPAS typically instead called SP) and the measured process variable $y(t)$ (in CPAS named, for example, PV or CV). The advantage with this form of the controller is that with scaled variables the controller gain K_C is dimensionless, while the other two parameters T_I and T_D have the unit time—and are consequently called integral time and derivative time, respectively.

As mentioned in the previous section, filtering the derivative part is important. Thus, the PID design should be considered a four-parameter design with the time constant of a filter T_F as the fourth tuning parameter. With filtered derivative action and using Laplace transforms, the controller output becomes

$$U(s) = K_C \left(1 + \frac{1}{T_I s} + \frac{T_D s}{T_F s + 1} \right) E(s)$$

Tuning Methods

There are many different principles based on which PID tuning methods have been developed, including the following:

- Tuning rules (Ziegler-Nichols, Cohen-Coon, etc.)
- Cancellation (IMC)
- Pole Placement
- Specification of Gain and/or Phase Margin
- Optimization

There is no possibility in this brief overview to go into the technical details on all of these concepts. Many of the classical tuning methods are covered in standard textbooks on process control; see, for example, Seborg et al. (2003). For a recent monograph on PID, see Åström and Hägglund (2006).

It can be concluded, though, that many methods lead to similar performance. Hence the differentiating factors on which to base a selection of a method are often factors other than pure control performance. Things to look out for are:

- Is there a tuning parameter? The user should preferably have a chance to make the trade-off between performance and robustness.
- Does it require numerical iteration, i.e., is there a risk that the method will fail to converge?
- Is the derivative filter part of the design?
- Are there any restrictions on the type of system treated?

Manual Tuning and Tuning Rules

As remarked above, in the beginning there was really only manual tuning. A typical procedure for manual tuning starts with using proportional only (P) control, increasing the gain until a stable self-oscillation in the process variable results. The gain is then decreased to about half this value and inte-

5.2 Loop Tuning

gral action is introduced. The original Ziegler-Nichols tuning rules used the same type of experimental procedure. When a sustained self-oscillation was achieved, the gain and oscillation period were noted and used to compute K_C and T_I (and T_D if PID was desired) from simple calculation rules. Published in 1942, these rules have now reached retirement age, but although they may have often led to overly aggressive control action they have served industry well. Meanwhile many more tuning rules, leading to better control performance, have been presented. Some build on the original oscillation experiment, but already Ziegler and Nichols themselves developed rules based on a step response experiment (also called a bump test), that is to say, when the controller output is subjected to a (typically manual) step change. A very comprehensive coverage of tuning rules is given in O'Dwyer (2006). See also Åström and Hägglund (2006) which includes the fairly recent κ-τ tuning rules which they put forward as "a Ziegler-Nichols replacement."

One advantage with tuning rules is that they require a minimum of support from the CPAS. The information required can often be determined from a trend plot or printout thereof. Another advantage is the relatively low requirements on the user, since a straightforward "cookbook" procedure can be written down. Disadvantages include a) that the experiment gives limited information about the process and b) a lack of flexibility. Since there is typically only one set of rules, there is no tuning parameter available to choose the level of aggressivity of the control.

Relay Auto-tuning

The original oscillation experiment has considerable disadvantages in that the user has no control over the amplitude of the oscillation created by the P control. An alternative experiment implemented in several commercial control systems was suggested by Åström and Hägglund in the 1980s; see their book (2006) for a more recent treatment. During the experiment the automatic feedback controller is replaced by a relay (potentially with hysteresis) (see Figure 5–7), which also leads to a self-oscillation of exactly the same character, but where the amplitude of the oscillation can be controlled by the selection of the amplitude of the relay. After determining the ultimate period P_u and the amplitude A, the PID parameters are calculated using (modified) Ziegler-Nichols rules.

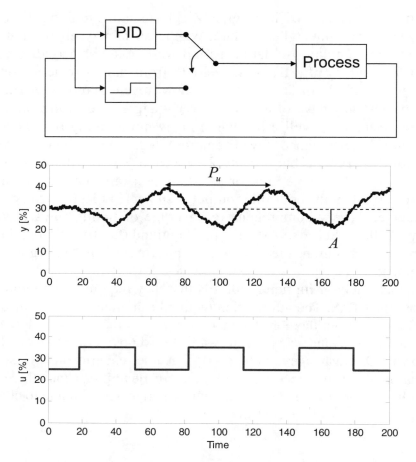

Figure 5–7 A Relay Auto-Tuning Experiment

Lambda Tuning

A popular method for tuning PI controllers is often referred to as Lambda Tuning (Thomasson, 1995). It is in fact two different methods; one for stable first order systems with delay and one for integrating systems.

Lambda Tuning—Stable Systems

The version for stable first order systems is based on a plant model which, expressed in terms of Laplace transforms, is given by the transfer function

5.2 Loop Tuning

$$G(s) = \frac{K}{Ts+1} e^{-sL}$$

The parameters of this so-called KLT model, K, L, and T, are known as the process gain, dead time, and time constant. An illustration of how they may be observed in the process variable when performing a bump test is shown in Figure 5–8.

The actual tuning method is based on selecting the integral time T_I such that the numerator of the controller cancels the denominator of the process model. The controller gain K_C is then chosen to obtain the desired closed-loop response determined by the desired time constant λ, approximately giving the closed-loop transfer function

$$G_{CL}(s) = \frac{1}{\lambda s + 1} e^{-sL}$$

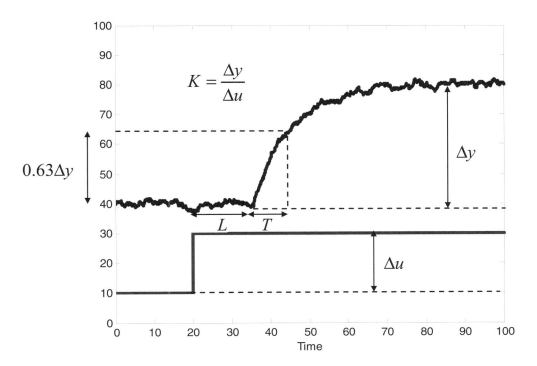

Figure 5–8   Step Response for a First Order Plus Dead Time System

Omitting the derivation, we conclude that the tuning parameters required may be calculated as

$$T_I = T$$
$$K_C = \frac{T}{K(\lambda + L)}$$

Lambda Tuning—Integrating Systems

Level-control processes, for example, have what is often referred to as an integrating response, i.e., when a step change is made to the controller output the process variable response is a ramp like the one depicted in Figure 5–9.

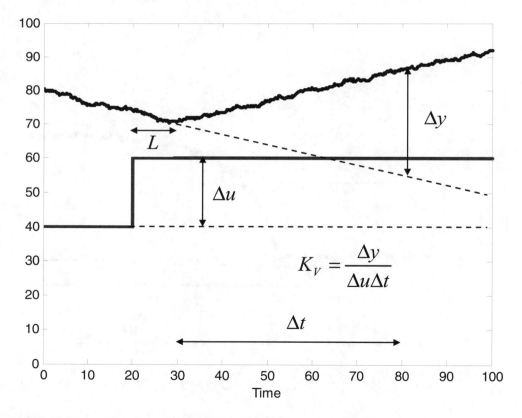

Figure 5–9 Step Response of First Order Integrating System

In terms of mathematics, this corresponds to the transfer function

$$G(s) = \frac{K_V}{s} e^{-sL}$$

Here cancellation cannot be applied and instead the controller parameters are derived which lead (approximately) to the closed-loop system

$$G_{CL}(s) = \frac{(2\lambda + L)s + 1}{(\lambda s + 1)^2} e^{-sL}$$

This is achieved by using the controller setting

$$T_I = 2\lambda + L$$
$$K_C = \frac{T_I}{K_V (\lambda + L)^2}$$

Lambda Tuning—Remarks

A few comments on the above presentation of Lambda tuning are called for. First of all, it is slightly unfortunate that the method for stable processes and the one for integrating processes are both referred to by the same name—Lambda tuning—since they are truly two different methods based on completely different control theory approaches, namely cancellation and pole placement, respectively. Just to emphasize this further, the first method does not tend to the second if $T \to \infty$, which you would otherwise expect, since then the stable system tends towards an integrating one. Skogestad (2003) presents a modified tuning rule:

$$T_I = \min[T, 4(\lambda + L)]$$
$$K_C = \frac{T}{K(\lambda + L)}$$

where, indeed, control of integrating process becomes a special case if one realizes that K_V corresponds to K/T.

However, the main advantage of Lambda tuning is its simplicity, and the fact that there is a single tuning parameter λ which has unit time. The main drawbacks are that the response may sometimes be unnecessarily sluggish, but more importantly, it cannot handle unstable systems. Nor is the method applicable to higher order models, but then there are straightforward extensions based on so-called internal model control (IMC) combined with model reduction; see Isaksson and Graebe (1999), Rivera and Morari (1987), and Skogestad (2003).

Model-based Interactive Tuning Tools

Even using relatively simple tuning methods like Lambda tuning, the user may benefit from an interactive software tool, for example to fit the model parameters to the step experiment data, or to interactively try out different values of λ and test them against simulations of the process.

Software support becomes necessary if the model is identified based on more general experiments than a single step. Also, for methods like Lambda tuning a more informative experiment—such as the one illustrated in Figure 5–10—may be advantageous. Over a longer experiment the noise may be averaged out, which is why a lower amplitude than for a single step may be used, thus disturbing production less.

Interactive tools are also essential when including derivative action since more than two parameters are difficult to tune manually or using simple tuning rules. As pointed out above, PID is really a four parameter controller. Unfortunately, most PID tuning methods do not consider the derivative filter as part of the design.

Two methods that do in fact optimize performance, taking all four parameters of the PID controller into consideration, are presented in Kristiansson and Lennartson (2006) and Garpinger (2009). However, none of them have so far made it into any commercial products.

Many distributed control systems have some built-in auto-tuning functionality in the PID controller, for example, relay auto-tuning. More rarely does a CPAS provide the user with possibilities for interactive tuning. On the other hand, most automation suppliers and other smaller companies sell stand-alone packages for loop tuning, which typically collect data from the process via OPC (one example, ABB's Loop Performance Manager is depicted in Figure 5–11). For an almost exhaustive list of different commercial PID controllers and tuning tools, see Li et al. (2006).

5.2 Loop Tuning

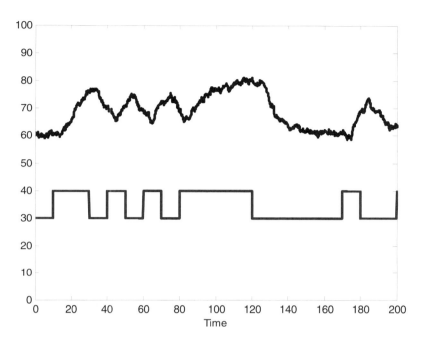

Figure 5–10   A More Informative Experiment with Multiple Steps of Varying Length

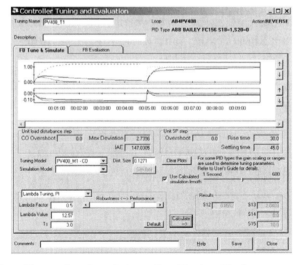

Figure 5–11   One Example of an Interactive Tuning Tool

REFERENCES

Åström, K.J, and Hägglund, T. (2006). *Advanced PID Control*. Research Triangle Park: ISA.

Garpinger, O. (2009). *Design of Robust PID Controllers with Constrained Control Signal Activity*. Licentiate Thesis, Lund University, Sweden.

Isaksson, A.J. and Graebe, S.F. (1999). "Analytical PID Parameter Expressions for Higher Order Systems." *Automatica* Vol. 35, No. 6, pp. 1121–1130.

Isaksson, A.J. and Graebe, S.F. (2002). "Derivative Filter is an Integral Part of PID Design." *IEE Proc.—Control Theory Appl.*, Vol. 149, No. 1, pp. 41–45.

Kristiansson, B. and Lennartson, B. (2006). "Robust Tuning of PI and PID Controllers." *IEEE Control Systems Magazine,* 26 (1), pp. 55–69.

Li, Y., Ang, K.H., and Chong, G.C.Y. (2006). "Patents, Software and Hardware for PID Control." *IEEE Control Systems Magazine*, 26 (1), pp. 42–54.

O'Dwyer, A. (2006). *Handbook of PI and PID Controller Tuning Rules, 2nd Edition*. London: Imperial College Press.

Rivera, D.E. and Morari, M. (1987). "Control Relevant Model Reduction Problems for SISO H_2, H_∞, and µ-Controller Synthesis." *Int. J. Control* 46(2), pp. 505–527.

Seborg, D.E., Edgar, T.F., and Mellichamp, D.A. (2003). *Process Dynamics and Control, 2nd Edition*. Hoboken: John Wiley & Sons.

Skogestad, S. (2003). "Simple Analytic Rules for Model Reduction and PID Controller Tuning." *Journal of Process Control*, Vol. 13, pp. 291–309.

Thomasson, F.Y. (1995). "Controller Tuning Methods." Sell, N.J. (editor), *Process Control Fundamentals for the Pulp and Paper Industry*. Norcross: TAPPI Press.

5.3 Loop Monitoring
A Key Factor for Operational Excellence

Alexander Horch

Motivation

Control loops are the heart of all production plants. They actively influence the production process such that it is able to keep physical measurement values close to their targets. Automatic control enables stabilization of inherently unstable processes, attenuation of disturbances, and reference tracking.

A critical prerequisite for advanced control, optimization, and other production improvements is properly functioning basic field-level control. It is therefore also the foundation of operational excellence and has to be one of the first issues to take care of when trying to achieve operational excellence in production.

Automatic control has been used successfully for decades, going from mechanical and pneumatical control via single-loop compact controllers to function blocks in distributed control systems. Modern production, however, requires faster production, faster change-overs to different products, better efficiency, and increased reliability. Control loops play a major role in achieving those objectives.

The assumption that good control is achieved once, when the plant is commissioned, and stays at that level during production is unfortunately often wrong. In many cases, control loop tuning is only done to such a level that the plant can be started successfully and is left in that state from then on. However, control loop performance is not static since the plant, operating conditions, and material change over time, resulting in control that is no longer optimal.

An automatic control loop, even if implemented in a modern CPAS, is not free. It has been estimated that each loop is approximately a $25,000 investment, including engineering, service, and hardware. Modern medium-size plants easily contain several hundreds or thousands of loops, corresponding to a multi-million-dollar investment.

It is reasonable to ask about the economic return from such an investment. It has been found that on many production sites, a considerable number (up to 30%) of control loops are constantly run in manual mode. The reason is that non-optimal tuning may increase process variability rather than attenuate disturbances. In those cases (if not set point tracking is a major goal) running loops in manual is better than automatic mode. This is a double loss of money: first, the investment was made but left unused; second, manual control usually corresponds to bad control performance, thus hindering operational excellence. Those control loops increase, rather than decrease, plant variability due to defects or bad tuning. Therefore, the potential of increasing performance is significant (decrease material and energy cost, increase production quality and throughput in the range of hundreds, thousands, or even millions of dollars), even if the percentage of non-optimal loops were "only" 5% for instance.

Loop Monitoring—As Old as Control Itself

In the same way as control performance is designed for in the initial phase, it is natural to ask if that same performance is still being achieved after some time of operation—and as we have seen, it often is not.

Early loop monitoring mostly used statistical process control methods. These were successfully used in manufacturing for a long time. A significant step towards practical monitoring of control loop performance was the work of Harris (1989), who suggested a simple-to-use performance index that measured the potential improvement of a loop compared to the best possible control (i.e., minimum-variance control).

This work initiated a large research and development effort that resulted in a plethora of performance monitoring methods and measures as well as commercial implementations thereof (VanDoren, 2008).

The methods developed during the last 15–20 years basically deal with the detection and diagnosis of two different aspects of controller performance: increased control loop variance and sustained control loop oscillations.

A control loop's inherent task is to keep the process value as close as possible to a desired target. The difference between these values is the control error, which should be kept as close to zero as possible. Obviously, any deviation from zero is undesirable. An oscillatory behavior of the control error usually represents a more severe problem than irregular variability. Therefore, development work has focused on detection and diagnosis of oscillatory behavior.

5.3 Loop Monitoring

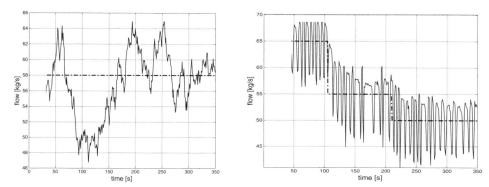

Figure 5–12   Control Error with Increased Variance (left); Control Error with Sustained Oscillation (right)

The second priority is to detect and diagnose irregular control error variations that can be removed by control. It was the contribution of Harris to develop a method to distinguish between variability that can be removed by feedback control and variability that cannot be removed from the process variable.

Figure 5–12 shows examples of loops exhibiting both general types of variability.

Among oscillation diagnosis research efforts, the detection of static friction (stiction) in control valves was once the predominant area of research. This diagnosis, as well as the detection of oscillations in general, can be considered solved nowadays. All modern software tools for automatic control loop performance monitoring offer such functionality. Current research and development mostly deal with the diagnosis of oscillations due to reasons other than stiction.

Loop Monitoring—Bottom-up Approach

Although a plant is controlled by many control loops, there is only a minority that has a major impact on plant economic performance. It seems that the Pareto principle also holds in this application. Identification of the critical 20% underperforming control loops will potentially remove 80% of the observed plant variability.

It is therefore important to focus on such loops. Usually, plant engineers have a good feeling as to which of the control loops are most critical for production and plant health. Then, by improving the most underperforming ones, it is expected that the observed plant variability will be reduced. Unless such a priorization is done, control loop monitoring will be done in a bottom-up fashion, i.e., monitoring more or less all control loops in the plant. Such an approach will inevitably find the underperforming 20%. The price is a larger effort required. The situation improves by adding a top-down approach which will be described below.

This section aims to present some of the metrics most often used to assess control loop performance on a single-loop basis. Since no dynamic model is usually known, the classic design metrics such as bandwidth or rise-time cannot be used in monitoring. This restriction is based on the fact that monitoring needs to be done with minimal information required. Hence, one must not require knowledge of a dynamic process model as is needed for controller design. The only information available is measurements of loop set point (the target), the process variable, and the manipulated variable (the controller output).

Unfortunately, there is a correlation between the simplicity of a measurement and its information value. Basic statistics such as mean and variance are easy to obtain, however, they do not give any indication of what the underlying problem could be.

Many different performance measurements have been proposed and are in industrial use. One important selection criterion for a performance measurement (index) should be if the information delivered can lead to a concrete action. Such an action could be a maintenance action such as re-tuning the controller, valve maintenance, sensor calibration, or process improvement. It could also be to scrutinize the loop in question since some problems cannot be detected and isolated without dedicated plant tests.

This chapter is too short to provide a comprehensive overview of all available monitoring methods. It aims instead to present certain categories of methods (with some examples) that may help the reader to assess available performance indices better.

Basic Statistics

Measures like mean and standard deviation, or higher-order statistics such as skewness and kurtosis, are simple and sometime useful for loop monitoring. Their basic drawback is that they do not represent an absolute value. A standard deviation of 5.3 may be "good" for one loop and "bad" for another. The value of such a basic statistic lies in tracking it. If such values are stored for a "good case," then their changes over time may well indicate some problem. However, neither the original values nor the changes will be able to tell the source of a problem.

5.3 Loop Monitoring 273

There are numerous other simple indices that may help an expert (or a system that represents expert knowledge) to assess loop performance. For example, assume that an oscillation has been detected. For the correct diagnosis of such an oscillation, it is of great importance to know its characteristics. These are, for instance, its frequency, its regularity, its symmetry, and its shape (sinusoidal, rectangular, triangular, irregular, etc.) Those values themselves may not permit a diagnosis, but they are important ingredients for more complex diagnostic algorithms and hypothesis testing.

Performance Measures

One important measure, the Harris Index (mentioned above) will be discussed further here. It is on a higher complexity level than the basic statistics above, and it allows a useful absolute interpretation.

The Harris Index allows an investigator to quantify the absolute improvement potential with respect to loop variability. Without showing the mathematics, the basic idea is the following: For a controller to remove a disturbance upsetting the process variable, it has to wait until the inherent process deadtime has elapsed. Dead-time is most often found in control loops in the process industry. Since the controller "has to wait," it is only able to counteract part of the incoming disturbance. The Harris Index relates the amount of variability that *can* be removed and the amount that *cannot*. Figure 5–13 shows an example: The Harris Index for the left picture is 2.12, indicating that the variance may be decreased by a factor of 2. The right picture shows the control error from the same control loop after re-tuning.

The performance index now is 1.04, indicating that the performance cannot become better, no matter which controller type or which controller tuning would be used.

Alternatives to Harris, and other performance measures, have been proposed and are available in commercial tools. Often they require careful use and significant data pre-processing. One such example is the Idle Index (Hägglund, 1999).

The Idle Index at time i (I_i) can be recursively computed from measured increments of controller output u and process variable y:

```
if ΔuΔy > 0, then s = 1
   else if ΔuΔy < 0, then s = -1
   else s = 0;
if s ≠ 0, then I_i = γ I_i + (1-γ) s
```

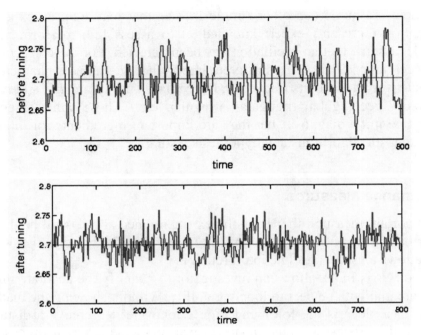

Figure 5–13   Control Error Before (top) and After Re-tuning (bottom)

The value γ determines the filter time horizon and is a function of the supervision time.

Oscillation Detection and Diagnosis

Oscillation detection, as simple as it is for the human eye, is not trivial to do automatically. However, as previously mentioned, it is a solved issue nowadays and available tools offer such functionality. It should be noted that pure detection (without any subsequent diagnosis) is of great value. A synopsis of all oscillatory loops in a plant may give great insight into reasons for performance degradation; see also the next section (Top-down Approach).

It was mentioned above that oscillations of control loops usually evolve from some root cause. The most prominent of such root causes are

- Aggressive controller tuning
- Nonlinear response problems in the final control element (most often valves)

5.3 Loop Monitoring

- External (oscillatory) disturbances
- Process nonlinearity
- … or any combination of the above

There is no single index or algorithm that can easily and safely diagnose any of these possible root causes. Much of the last decade's research has focused on this problem and some very promising results are already available for use in CPASs.

A key to solving this problem is to measure the degree of nonlinearity that can be found in the measurement data. This can be done by analyzing the shape of the oscillation (Horch, 1999) or by quantifying the inherent nonlinearity, e.g., by use of higher-order statistics (Thornhill, 2005). As an example, consider Figure 5–14, which shows two oscillatory data sets that exhibit stiction and an external disturbance, respectively.

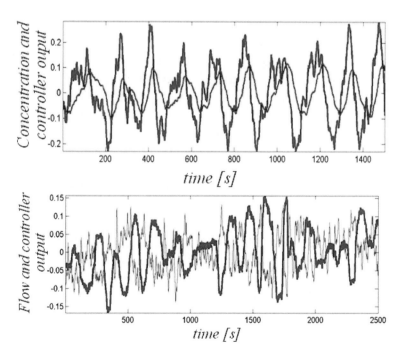

Figure 5–14 Process Variable and Controller Output for a Concentration Control Loop that Exhibits Stiction (top) and a Flow Control Loop that is Influenced by that Loop (bottom)

A specific problem for these advanced methods is the fact that industrial measurement data may contain all kinds of artifacts. They may have natural explanations, however, for automatic monitoring methods this is a big challenge. It is therefore important that the monitoring functions have access to additional information about the plant. In power plants, control loops are often responsible for a certain operating range. If a process variable leaves this range, another controller may take over. In such a case, of course, the original controller must not be blamed for deteriorated control since it is not in charge at that specific point of time. In a CPAS, there is easy access to information about control loop or process statuses and they can be combined in order to avoid erroneous diagnoses.

Loop Monitoring—Top-down Approach

Another approach to loop monitoring does not scrutinize each and every loop homogeneously. Instead, a high-level plant view is taken. In this approach, an individual control loop will only be investigated if it is pointed out as a potential root cause of a plant-wide disturbance. A recent overview of this field of research has been published by Thornhill and Horch (2007).

Easy and flexible access to historical data in a CPAS easily allows the presentation of plant-wide measurement data as shown in Figure 5–15.

This view reveals much more information than the usual operator trends, which have another purpose. The plant-wide disturbance analysis methodology uses such time-series plots (or rather their signal spectra) in order to identify clusters of data sharing similar disturbance patterns, e.g., sustained oscillations such as the ones above.

By using advanced statistical signal processing, it is then possible to identify those measurement tags that—statistically—are assumed to have influenced the other tags more than the other way around.

Application of such a technique to the data set in Figure 5–15 revealed that loop QC193 is close to the oscillation root cause. This result cannot be identified by visual inspection at all. All the other oscillatory signals (control loops) more or less share the same oscillation frequency since they are influenced by QC193.

A single-loop monitoring procedure of QC193 revealed that the loop had a nonlinear problem (dead-zone) that caused the oscillation. As a consequence, other loops across the plant were affected too.

5.3 Loop Monitoring

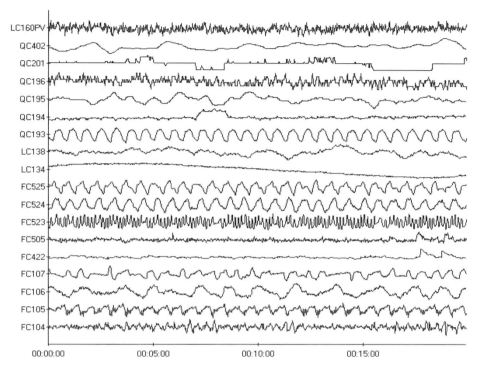

Figure 5–15 Overview of Plant-wide Measurement Data

Here the advantage of the top-down approach is obvious: It is not necessary to investigate every control loop. It is sufficient to concentrate on the one that has been pointed out by the plant-wide disturbance analysis result.

Figure 5–16 shows a simple schematic of the underlying process, a stock preparation section in a paper mill.

Plant-wide disturbance analysis will never be completely automatic, but it can be of great help to the plant engineer as it takes over considerable parts of the daily detective work. The result of such work is shown in Figure 5–17.

Implementation Aspects

The methods described above share one common property: they all analyze measurement data (whether on-line or off-line) in order to offer decision support. In a CPAS, such information is available anywhere in the system.

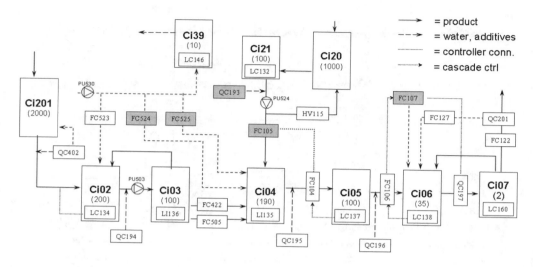

Figure 5–16   Schematic of Where the Data in Figure 5–15 Originates

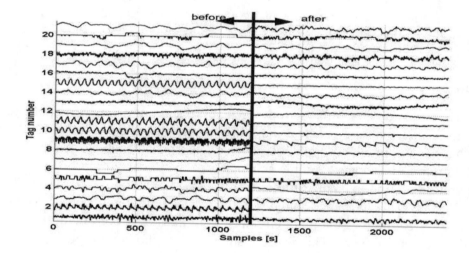

Figure 5–17   Data from the Plant in Figure 5–16 Before and After the Maintenance of QC193 (Tag No. 15)

Therefore, the implementation of loop monitoring (bottom-up on the single-loop level or top-down on the plant level) can be designed very flexibly.

Available implementations can be found in the controller hardware as function blocks. Simple monitoring functionality can be offered as properties within the control system in the same way that trends or documentation may be

5.3 *Loop Monitoring*

accessed. Loop monitoring can be an integral part of the Asset Management function of the CPAS. Or, if desired, it can be performed completely disconnected and on separate hardware, being connected to a historian via OPC only. Such a tool offers in-depth analysis capability on top of standard monitoring. Finally, basic loop assessment can also be performed directly, within the data historian.

In order to give a concrete example of an implementation, see Figure 5–18.

Conclusion

Loop monitoring has become an integral part of modern plant management, especially within initiatives that focus on operational excellence. Most advanced applications rely on solid basic control. Even though the most prominent industrial problems can be diagnosed nowadays, there is still much difference in effort required, quality, and interpretation of monitoring results, depending on the vendor of a monitoring software package.

Loop monitoring has become a standard requirement for many modern production plants and has been included in standard maintenance programs.

There are no standards at present, such that each user is able to find out what exactly he or she wants to know about the plant's control performance and how to acquire it. Much of the required technology, however, is already available and offers great flexibility. This flexibility is provided in large part by the free access to high-quality measurement data in a modern CPAS.

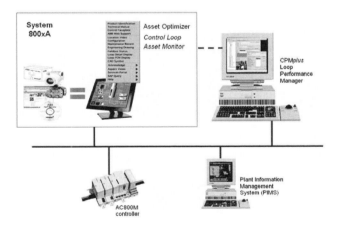

Figure 5–18   Possible Implementations of Loop Monitoring in CPAS

References

Harris, T. "Assessment of Control Loop Performance." *The Canadian Journal of Chemical Engineering* 67 (1989) pp. 856–861.

VanDoren, V. "Advances in Control Loop Optimization." *Control Engineering* 5/1/2008 (http://www.controleng.com/article/CA6559117.html).

Hägglund, T. "Automatic Detection of Sluggish Control Loops." *Control Engineering Practice* 7 (1999) pp. 1505–1511.

Thornhill, N.F. "Finding the Source of Nonlinearity in a Process with Plant-wide Oscillation." *IEEE Transactions on Control System Technology* 13 (2005) pp. 434–443.

Horch, A. "A Simple Method for Detection of Stiction in Process Control Loops." *Control Engineering Practice* 7 (10) (1999) pp. 1221–1231.

Thornhill, N.F. and Horch, A. "Advances and New Directions in Plant-wide Disturbance Detection and Diagnosis." *Control Engineering Practice* 15 (2007) pp. 1196–1206.

5.4 Plant Asset Management
Plant Asset Health—A Key Factor for Plant Management

Alexander Horch and Margret Bauer

Toward a Common Understanding

The term Asset Management originates from the financial markets, where people strive to maximize the return on investment in (financial) assets. Therefore, the term Asset Optimization is also used by some vendors. In industrial production, any investment aims for some kind of return maximization. It was therefore natural to use the same expression there as well.

In order to distinguish *industrial* asset management from *financial*, *digital*, or *information* asset management or others, it has recently been proposed to use the term "Plant Asset Management" (VDI/VDE, 2008). It refers specifically to production assets in production plants and mainly focuses on the handling of such assets during the operation and maintenance phases of the asset life cycle: *"Assets of a production plant are all components that are directly required for production... Excluded are infrastructure, such as roads, and human resources.* (VDI/VDE, 2008).

In general, "(plant) asset management" and "(plant) asset optimization" denote similar tasks and objectives; however, "optimization" suggests a more value-driven approach than pure "management" (Reierson, 2006). In practice, this seems to be a more philosophical question, although it may be useful to think of asset optimization as a core value or objective of asset management.

A possible categorization of plant assets is proposed as follows:

- Assets on the plant level (e.g., an ethylene plant)
- Assets on plant section level (e.g., switch gear, compressor station)

- Assets on plant component level
    - Static equipment (e.g., tank, pipe, heat exchanger)
    - Rotating equipment (e.g., pump, compressor)
    - Machines (e.g., packaging machine)
    - Automation field devices (e.g., valve, motor)
    - Automation hardware (e.g., communication network)
    - Automation software (e.g., process model, control function)
    - Miscellaneous components

The term asset management (and its alternatives) have often been loosely defined, which has led to completely different understandings. In VDI/VDE (2008), a definition is proposed that helps in understanding what the essence of asset management is:

- Management of assets during their entire life cycle. The main focus covers identification, asset history, and economic and technical data.
- Organization of the deployment and preservation of assets.
- Generation and provision of information, especially concerning trends and prognoses about asset health for decision support.

The above asset management activities have certain optimization objectives in relation to process plants:

- Attaining the best reliability and efficiency of the assets.
- Improving each asset's value by enhancing the application and minimizing maintenance costs.
- Reducing replacement demands by optimum application and best possible preservation of the existing assets.

A major misunderstanding about asset management is that people feel the need to discuss whether or not they "need asset management." In fact, any production plant is doing some kind of asset management. They may or may not, however, have the tools and methods to support it efficiently.

5.4 Plant Asset Management

This becomes clear from the definition of asset management. Each plant manager tries to use the production facility as best as he or she is able, given the specific requirements.

Therefore, asset management deals with certain tasks in order to achieve the objectives given above. A CPAS offers functionality and infrastructure in order to execute such plant asset management tasks.

Figure 5–19 gives a short list of the most important of such tasks (VDI/VDE, 2008).

Which Maintenance Strategy?

Plant maintenance represents the closest link to plant asset management. Maintenance staff both uses and delivers information for an asset management system. Many of the tasks of plant asset management are done by maintenance staff; however, asset management is much broader and therefore touches many other user groups such as operations, engineering, and plant management.

PAM tasks	
Monitoring, Diagnosis, Prognosis, Therapy	
Performance Monitoring	
	Plant Performance Monitoring
	Unit and Component Performance Monitoring
	Production Performance Monitoring
	Control Performance Monitoring
Management of asset alerts	
Condition Monitoring	
	Signal monitoring
	Function test
	Diagnosis
	Collection of maintenance requests
	Condition Prognosis
	Estimation of remaining life-time
	Proposal of Therapy
Performance Optimisation	
Information allocation and archiving	
Document management	
Asset history	
Version management	
Analysis, Representation, Distribution	
Management of user action (Audit Trail)	
Calibration and gauging	
Configuration and calibration	

Figure 5–19 Plant Asset Management Tasks

Every company runs a different maintenance strategy and plant asset management does not force a certain maintenance strategy, but rather supports the current maintenance strategy that is pursued in a plant.

There are four classes of maintenance strategy under the plant's general strategy, and most companies execute a mixture (or just one, e.g., reactive maintenance) of all four of them in a different distribution. Table 5–1 shows these four different strategies and a recommendation for a distribution that can be found in production sites with operational excellence.[1]

Each of these maintenance strategies is appropriate in certain cases. Non-critical, cheap equipment with low risk of failure should be run reactively, whereas critical equipment with a high impact on productivity should be maintained proactively or predictively.

As mentioned, most plants are maintained using a different mix of the above strategies. In general, it can be stated that many companies still lag behind the recommended mix of reactive, preventive and predictive maintenance. There is, however, no absolute recommendation since the maintenance strategy should be a consequence of the overall company strategy. This needs to be seen in the context of the business in general and may look different between two companies, both having chosen an optimal mix.

In any case, it is important that for each asset, a maintenance strategy is chosen that fits into the overall company strategy as well as being related to probability of failure and its potential impact on production (see Figure 5–20).

A major task of an asset management system is to support the plant's maintenance strategies such that the objectives outlined in the previous section will be realized.

Which Assets Should be Optimized?

It has been pointed out that all assets directly related to production need to be the object of asset management. In the previous section, it was stated that critical assets need to be maintained using a more appropriate maintenance strategy (e.g., preventive rather than reactive or predictive rather than preventive). It is obvious that for the successful realization of an asset management system, the first effort should be directed towards the critical assets.

---

1. Source: Tooling University, 15700 S. Waterloo Rd., Cleveland, OH 44110-3898

5.4 Plant Asset Management

Table 5–1 General Maintenance Strategies

Strategy	What?	Typical Usage	Goal	Definition
Reactive	"Run till it breaks"	> 55%	10%	Maintenance performed only after a machine fails or experiences problems.
Preventive	"On schedule"	31%	25–35%	Maintenance performed while a machine is in working order to keep it from breaking down.
Predictive	"Only when required"	12%	45–55%	A maintenance approach that involves testing and monitoring machines in order to predict and prevent machine failures.
Proactive	"Eliminate root-causes"	~2%	Rest	Maintenance is based on root cause analysis results that detect disturbances in the process.

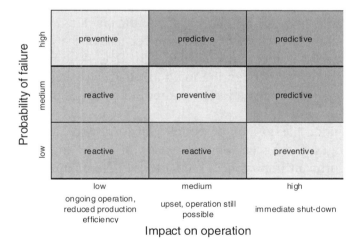

Figure 5–20 Reasonable Maintenance Strategy Based on Probability of Failure and Impact on Operation per Asset

In many industrial applications, the implementation strategy chosen is different. Often, the strategy is first applied at the field device level, using asset management for device management and configuration. This strategy follows the development of asset management itself, which started with field devices in the first place, see also Muller, J., Gote, M., and DeLeeuw, V. 2007.

There is nothing contradicting such a strategy; however, when efforts stop at this point, the benefits will be hard to prove since field devices most often do not represent the most critical process equipment.

Another area of application is condition monitoring of critical equipment. These often expensive and resource-intensive solutions are often applied to equipment that is felt to require predictive strategies.

In practice, however, such solutions are most often isolated applications that are not interconnected with others or with a central intelligence. As a consequence, the expected benefit often cannot be fully realized.

Here is the point where the broader topic of "optimization" comes in. As has been stated, plant maintenance represents the closest link to plant asset management, but asset management is much broader and therefore touches many other user groups such as operations, engineering, and plant management. When realizing plant asset management solutions, the first step is to define an asset management strategy that relates to the business strategy of the company. This means it has to be supported strongly by management. Without such support, most such efforts are not successful.

When the strategic dimension is sorted out, an implementation plan has to be developed. This plan needs to be specific to the plant in question, first because of the specific company strategy and second because of the specific nature of the plant. A matrix similar in principle to Figure 5–20 needs to be identified.

When it is clear which are the main assets to be taken into consideration and what level of implementation is desired, one may start to plan a concrete realization.

Such planning has to include people, infrastructure, and software. In this book, the changes in the work environment will not be discussed even though they are of significant importance, see Gote, M., Neumann, S., Bauer, M., and Horch 2008.

By its nature, asset management is an integral part of a CPAS. A central requirement for an asset management system is a centralized information portal that collects, analyzes, distributes, and communicates information to the different user groups. An abstract representation of such an asset data portal was also proposed in VDI/VDE (2008); see Figure 5–21.

It should be pointed out that this model is not a building plan for an asset management system. Its purpose is to foster common understanding of plant asset management.

It is important to understand that the building blocks of the model in Figure 5–21 can have different realizations. The maintenance technician put-

5.4 Plant Asset Management

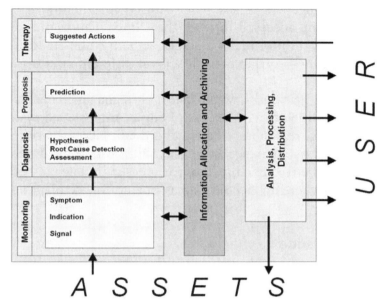

Figure 5–21 Abstract Plant Asset Management (PAM) Model (Describes the functional and logical dependencies for the execution of asset management tasks. The task blocks that support plant asset management are pointed out, as well as the required external interfaces to users and assets.)

ting his or her hand on a motor performs the same basic task as a temperature sensor or a piece of complex monitoring equipment. There is currently no standard that suggests where, how, and if (at all) to realize the different specific functions.

Within the CPAS paradigm, however, it is obvious that a system underlying the collaborative character of automation may greatly enable and improve a holistic approach to plant asset management.

Asset Management and Maintenance

It was pointed out before that plant asset management needs to be implemented in a holistic manner, which covers the maintenance and operation workforce as well as an overall plant production strategy.

In the following, the technical aspect of plant asset management will be put into focus. The collaborative nature of a CPAS represents a perfect basis for plant asset management tasks. All required information as well as all potential users can be accessed through the CPAS. This avoids some of the usual problems: lack of information, lack of actuality, non-standardized information, the user being forced to consult several different systems for decision support.

There are many use cases within plant asset management, out of which some will be exemplified here. The task of engineering such a system will not be covered; it is, however, an important and difficult piece of work.

Three different maintenance use cases will be presented here: reactive, preventive, and predictive. Each of those use cases initiates a work order for maintenance and—where applicable—an asset alert for the plant operator.

Reactive Maintenance

As mentioned, this maintenance approach is appropriately applied to non-critical assets. Their failure does not harm production or result in extensive cost. However, when equipment fails, it is important that the staff in charge of that equipment can identify such a failure easy and quickly.

Asset health information, including consequence analysis and suggested maintenance action, needs to be conveyed to the maintenance staff, whereas the operator needs to be informed only if the failure would influence current or future operation of the plant.

Preventive Maintenance

Preventive maintenance is the predominant maintenance strategy in many production plants. The scheduled maintenance actions are often the result of intensive optimization based on good experience.

Scheduled maintenance actions are mostly stored in the computerized maintenance management system (CMMS) which is the main software tool the maintenance staff is working with.

For plant asset management to work well, a good integration of a CPAS with such systems is a key requirement. How such an integration may be successful is further described in Section 5.6.

Having such an integration in place will enable the displaying of scheduled maintenance actions to the operator (in case he or she is affected) and to the maintenance staff (who has to perform the action).

5.4 Plant Asset Management

A modern asset management system should offer the ability to map scheduled maintenance tasks to actually required maintenance tasks, thus enabling optimization of the preventive maintenance schedule.

Predictive Maintenance

This maintenance strategy is most often referred to in connection with plant asset management. It is the most difficult and most beneficial aspect of asset management. By knowing the asset health of critical process equipment in advance, much of the potential benefit of plant asset management may be realized.

Knowing in advance that things will need to be maintained during a certain period of time enables the plant operations team to make changes in plant operation, e.g., re-scheduling production, running at a more plant-friendly operating point.

The maintenance staff may re-schedule maintenance tasks in order to run the high-priority tasks first, maximizing plant availability and minimizing shutdowns.

The distribution of asset alerts within a CPAS follows modern standards: email, pager, alarm messages, etc. One important aspect of such a functionality is that the asset health information needs to be in a consistent and unified format. If asset information is conveyed in different formats with different meanings by different sub-systems, no aggregation and unified data handling will be possible.

Realization of Plant Asset Management

As an example of how plant asset management might be implemented, consider the CPAS diagrammed in Figure 5–22. In order to treat asset information from different assets (plant section or simple field device) in the same way, it is necessary to publish asset health information in a standardized way.

Since there are currently no accepted standards for such information representation, some CPAS suppliers have chosen to transfer any asset health information into a so-called asset condition document. Based on such a document, all other CPAS functionality is developed (see Figure 5–22).

Referring to the PAM model (Figure 5–21), the left part is realized using asset monitors. Asset monitors are small data processing units that either collect information from decentralized intelligence (e.g., field device

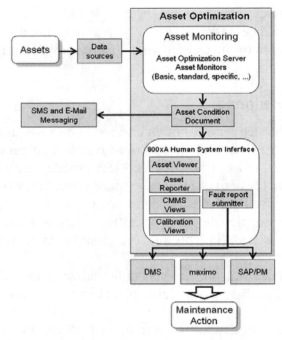

Figure 5–22 Management of Asset Health Information in a CPAS

diagnostic information) or generate asset health information themselves by collecting and aggregating other information.

An asset monitor (Figure 5–23) reads any data as input and generates an output that is standardized within the CPAS. The input information may be diagnostic information from third-party systems, field device information, measurement values, alarm messages, database entries and the like.

Whenever asset health information is available in the asset monitor, it can be used anywhere within the CPAS. Due to the object-oriented nature of the system, it can be accessed from any plant object, e.g., from an alarm message or from the graphical object in the user screen.

Asset health messages can be conveyed in many ways. To the extent the operator needs to know about asset health, information may be presented on the operator screen. Maintenance staff can be alerted via asset health events in a user screen or any messaging service.

Due to the central administration of asset health data, an overview of complete plant health can be gained very easily; see Figure 5–24 for an example.

It should be noted here that a modern CPAS offers freedom to realize asset intelligence wherever the current asset requires it. One extreme is a field

5.4 Plant Asset Management

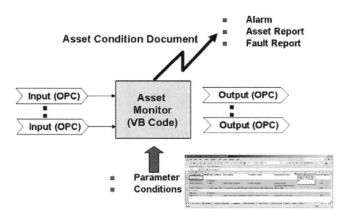

Figure 5–23 General Structure of an Asset Monitor

device that includes a decentralized unit which is able to deliver complete asset health information. The other extreme is feedback from a plant inspection performed by a human. Both types of information enter the CPAS in different ways; however, within the system it is handled equally, thus enabling real information aggregation. It is only by this that a real plant-wide asset health assessment becomes possible.

Conclusion

It has briefly been shown which ingredients an asset management system that is realized as part of a CPAS will contain. Plant asset management is still very much a misunderstood area of application. However, great steps forward have already been made and solutions for significant progress are already available.

A lack of general understanding, standards, and applications still hinders the broad use of automated plant asset management solutions in industry.

A CPAS and its possibilities offer the main missing link between different—and already accepted—applications and will greatly help in realizing the expected benefit of plant asset management in modern production.

As a final example, consider the case for monitoring a medium-voltage drive (Figure 5–25). Such a package unit may serve as a "mini-plant" where both the application and benefit of plant asset management are already industrial practice. This example shows that plant asset management may start on a modular and very well specified level. When having identified critical assets, it is crucial for success that those assets can be monitored without installing large

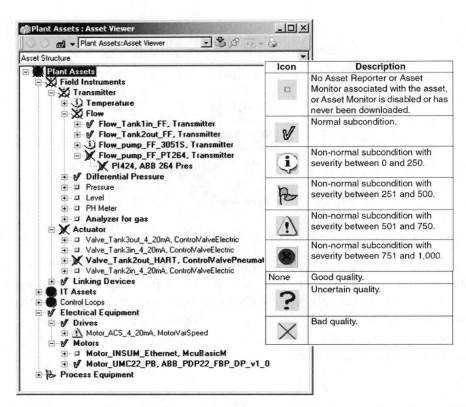

Figure 5–24 Example of a Plant Section Tree View (with summary information about all assets that are being monitored within that specific plant section.)

systems. Having such a local system in place then is the basis for extension to other assets as required.

Figure 5–25 shows the maintenance screen that tells the maintenance technician about the status of the drive and where problems arise and what to do in such case.

References

Reierson, B. "Myths and Realities of Asset Optimization." *Plant Engineering*, April 2006 (www.plantengineering.com/article/CA6308016.html).

5.4 *Plant Asset Management* 293

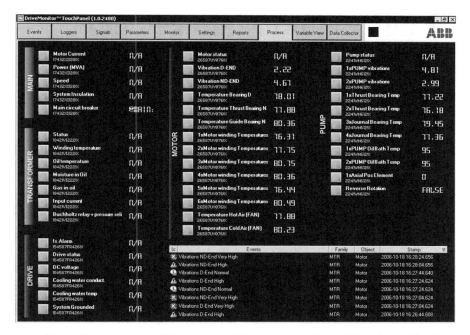

Figure 5–25 Example of a Plant Asset Management User View for a Medium-Voltage Drive

Gote, M., Neumann, S., Bauer, M., and Horch, A. "Trends in Operations and Plant Asset Management—A Discussion Contribution." *Automatisierungstechnische Praxis Atp*, 10/08 pp. 48–54.

VDI/VDE "Plant Asset Management (PAM) in the Process Industry—Definition, Model, Task, Benefit." *VDI/VDE Guideline* No. 2651, 2008.

Müller, J., Gote, M., and DeLeeuw, V. "NAMUR Umfrage Plant Asset Management" *Automatisierungstechnische Praxis Atp,* 01/2007.

5.5 Information Management

Martin Hollender

Introduction

Information is a key ingredient in achieving a sustainable competitive advantage. Reduced time to decision and action is critical for improving quality and productivity. This makes the timely collection, transformation, and distribution of reliable process information a significant issue. One of the barriers to increased productivity is aggregating data from a myriad of disparate sources, transforming it into meaningful information, and presenting it to operations, maintenance, engineering, and management in the context most meaningful to them.

In the past, a DCS (Distributed Control System) usually had only limited Information Management capabilities and if data storage was possible at all, the storage horizon was less than a few days. A separate Plant Information Management System (PIMS) had to be added for Information Management functionality. In today's systems, Information Management has become so essential that it has become logically integrated as a standard part of a CPAS. Nonetheless, a clear physical separation of core DCS and Information Management functionality still makes sense even in a modern CPAS, as will be explained later in this chapter.

An open distributed architecture supports data storage in either a multiple history server consolidation in one location, or in several locations for additional fault tolerance. Information Management needs to have near-real-time characteristics. It is not necessary to respond to information requests as fast as automation controllers do (milliseconds to seconds), but depending on the application, Information Management needs to be able to respond in the seconds to minutes range. For example, some Advanced Process Control applications as described in Section 5.1 can be built on top of Information Management. A plant manager using Information Management usually wants to know what has happened in the last month, week, shift, hour or even minutes.

5.5 Information Management

In many cases a detailed record of production is required by law. The Food and Drug Administration (FDA) regulation 21 CFR Part 11 covers electronic records and signatures that affect the production, quality, and distribution of drugs in pharmaceutical units (see Section 2.9 on traceability).

In case of a process upset, an incident, or an accident, Information Management is an important basis for the analysis of the event. In such cases the Information Management module also acts as a "flight recorder" (an analogy to the black boxes installed in aircraft) for the plant. Therefore, an Information Management module is becoming more and more mandatory in safety-critical plants such as petrochemical operations.

In addition, the Information Management infrastructure can be used as the basis for production and asset optimization, machinery monitoring, raw material utilization, batch tracking, and many other applications.

The thick clients shown in Figure 5–26 are classical applications that need to be installed on the client computer. A typical example for a thick client is Microsoft Excel. Thick client applications can offer a high-quality user interface and can use the full spectrum of functionality available on the client operating system. Disadvantages of thick client technology are the high cost for deployment, maintenance and updates. Thin client technology makes the results of Information Management applications available via standard Web Browsers (see the section on Wide Area Distribution in this chapter).

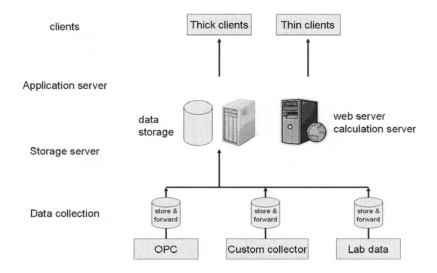

Figure 5–26 Information Management Architecture

Key Information Management Requirements

Of course, there is a large number of requirements for Information Management. In the following, the most important of these requirements are discussed.

Separation of Information Management from Core Production Control

Information Management applications need to be clearly separated from core production control because they should never interfere. Keeping the process stable is usually the most important priority. Interference from Information Management applications might happen for the following reasons:

- In some cases, applications might request large amounts of high-resolution process data. For example, a Data Reconciliation application might subscribe to several ten thousand process measurements. It is an important task of the Information Management component to throttle down such requests to rates that the core production control system can handle without being disturbed.

- An Information Management application might be very demanding and might use 100% of the available resources. In such a case it is important that the application runs on a separate server, because otherwise it might disturb functions of the core production control system. For example an unconstrained query in an Alarm Management report might involve several millions of events and therefore consume significant resources.

In addition, Information Management is usually the gateway to the world outside of the control room so that offices like the plant management or the corporate headquarters can access plant data. From a security point of view, it is better to have such a gateway in a separated network zone (see Section 2.8 on security). Somebody from within the intranet with access to Information Management might (accidentally or intentionally) issue bogus queries or calculation requests that create a very heavy load. If Information Management and core production control are too tightly coupled, such a denial-of-service attack might even stop production. When the systems are clearly separated, the attack would only disturb Information Management, which is not directly needed for ongoing production.

5.5 Information Management

Information Management applications are usually installed on separate servers with a well-defined interface to core production control. This ensures the required separation. As typical Information Management applications and core production control have different requirements concerning availability and robustness, a CPAS stores the historical values for operator trends used for core production control (with an hours to days horizon) in a separate system. The more advanced functionality of Information Management (such as reports and calculations) is important for the control room also, but here the requirements are less strict.

Fault Tolerance

Computers might crash or be unavailable and network connections might be cut. Data acquisition still needs to work without losing any data. This can be achieved with two strategies:

- Redundant data collectors and servers. If one of the servers or its connected data collector fails, the redundant server/collector pair takes over and fills back the data once the failed server and its data collector are operational again.

- Store-and-forward data collectors. The data collector buffers data locally and waits to delete the buffer until it gets a return receipt from the server that the data was successfully stored. In case of a disturbed connection or if the receiving server is unavailable, the data remains on the data collector. Once the receiving server signals its availability again, the buffered data can be transferred. This scheme is sometimes used with a non-redundant server assuming that a failed server can be quickly fixed and then be back filled from a data collector.

The Information Management servers need to be supervised, so that problems can be discovered early and can be fixed promptly. A loss of production data can have dramatic consequences, especially if legal requirements are involved.

Data Acquisition from Heterogeneous Sources

Information Management needs to consolidate all data relevant for operation in one unified view. Often this is not an easy task as many plants have evolved over time and use automation technology of different generations and from different vendors. Other sources such as Laboratory Information Management Systems (LIMS) and Enterprise Resource Planning (ERP) have important information that needs to be integrated with process data to get a more complete picture. If modern acquisition interfaces like OPC are not available, Information Management should provide custom drivers for the most common systems. Otherwise, a project-specific driver needs to be implemented.

Information Management Modules

As shown in Figure 5–26, the main components of Information Management are collection of data, data storage, calculation/application server and information distribution. In the following, each of the modules is described.

Data Acquisition Interfaces

The state-of-the-art data acquisition interfaces are OPC DA for real-time process values and OPC AE for alarms and events (see Section 2.3).

But as one function of Information Management is a consolidated view of the overall system, the ability to integrate all kinds of external sources like legacy DCS is very important. Such interfaces are often developed on a per project basis, and the development can be very expensive. A library of configurable data acquisition interfaces to the most common external information sources without standard OPC interfaces is therefore of great value.

Other possibilities include reading comma-separated values (CSV) or similar intermediate formats from a directory. If alarms and events are not available from OPC AE, they are usually recorded from a standard printer port, as printing alarms on an alarm printer was common for older systems.

Long-term Data Storage

The requirements for storing process data range from a few weeks to having more than 10 years of data available online. Data storage is important because:

- In some industries storage of production data is required by law. If, for example, pharmaceutical production records get lost, the corresponding

5.5 Information Management 299

drugs must be thrown away. Power plants must prove that their emissions are according to their bought allowances from emissions trading.

- Historical process data can be compared with current production data to be able to discover trends and degradations.

The price for disk storage has dramatically decreased during the last years. One Terabyte of redundant array of independent disks (RAID) storage is a commodity today that can be bought in a shop around the corner. In addition, the availability of cheap, reliable, and fast Internet connections together with cheap and reliable data storage and computing power in commercial data centers might allow outsourcing parts of Information Management (such as data storage) to external service providers in the future.

Process Value Storage and Compression

A process value recorded uncompressed, once per second with 16-bit resolution, would need more than 60 MB per year. This would have the following disadvantages:

- Even though disk space is rapidly getting cheaper, disk cost still would be significant for a typical plant with 10,000 or more stored values. Fast processes like steel cold-rolling mills need to sample data many times per second.

- Speed of data access is very important, e.g., when opening a trend display, and disk access is an important limiting factor because disk access is comparably slow.

- Many applications don't require high resolution data. For example, a trend showing last year's data cannot show more pixels than configured in the display. It would be a waste of network resources to send millions of values if the screen does not allow showing more than 1280 different values.

- Many of the measured values are not changing very fast, and are therefore good candidates for compression.

Several different strategies and methods are available to compress sequential process values:

- One widely used method is using hierarchical structures. High resolution data is only stored for a short period of time, perhaps the last month. For periods longer than that, only some condensed values are stored, e.g., 10 minutes average and/or 10 minutes maximum values. These hierarchical logs have dramatically lower disk storage requirements.

- As many process variables change rarely, the use of deadband algorithms can drastically reduce the required storage space. Such algorithms place a deadband around the current value and store the next value only when the process variable leaves the deadband. As the size of the deadband is usually calculated with the help of the measurement range, these ranges need to be correctly specified, because otherwise the deadband becomes too wide and the accuracy of the data becomes too low for many applications (see Thornhill et al., 2004).

- The swinging door algorithm (Bristol, 1990), stores a value if a straight line drawn between the last stored value and the next value does not come within the compression deviation specification of all the intermediate points. The compression deviation needs to be carefully engineered, because if the tolerance band is selected too wide, no values will be stored even if interesting changes have occurred.

- Run-length encoding is a simple form of data compression in which runs of data (that is, sequences in which the same data value occurs in many consecutive data elements) are stored as a single data value and count. As most of the bits are not changing from step to step, run-length encoding is a fast and efficient compression method for process data.

- Other compression techniques include zip or wavelet compression (Watson et al., 1995).

The compression methods used for Information Management are usually lousy, which means that the original high-resolution data cannot be fully reconstructed from the compressed data. The compression parameters need to be manually engineered and as a consequence, the parameters are often left at their default values, typically 1% of the maximum value. This is often larger than the actual noise variance in the data, resulting in a significant loss of information (Alsmeyer, 2006).

5.5 Information Management

Alarms & Event Storage

Alarm and events have fields like timestamp, tag name, message, and severity, and are very well suited for storage in standard databases (Figure 5–27). One common problem is that some rapidly recurring nuisance message might rapidly fill the database and overwrite important messages. It is therefore very important to be able to filter out such nuisance messages before they reach the database.

The OPC AE standard allows specifying vendor-specific attributes. Each OPC AE server can create its own vendor-specific attributes. CPASs often consolidate large numbers of underlying OPC AE servers (see Section 4.1). The resulting systems therefore need to manage a large number of different vendor-specific attributes.

It can happen that two different OPC AE event sources use different terms for the same thing. For example, one server may define an attribute "nodename" and another attribute "server," both meaning the machine where the event was generated. If later on an application needs to evaluate all events generated on a specific machine, it is likely that one of the attributes will be overlooked. Therefore, Information Management needs to harmonize the vendor-specific attributes and store semantically identical information in the same attribute.

Calculations

One possibility for the execution of algorithms is to execute them immediately after the acquisition of the data before it is stored. Standard algorithms include maximum, minimum, average, sum, and sum of squares values for a specified time interval. Another possibility is to run calculations at scheduled intervals, e.g., every minute or every day, to compute derived values. These derived values can be stored in addition to the values directly measured from the process. Example of calculations include:

- Statistical Analysis for Quality Control and Performance Monitoring.
- Boiler Lifetime Monitoring. Component stresses in steam generators are strongly influenced by start-up and shut-down operations as well as varying load cycles in the form of varying steam pressures and high thermal stresses. The varying loads lead to cyclic strain exhaustion. Process data such as pressures, metal temperatures, and temperature

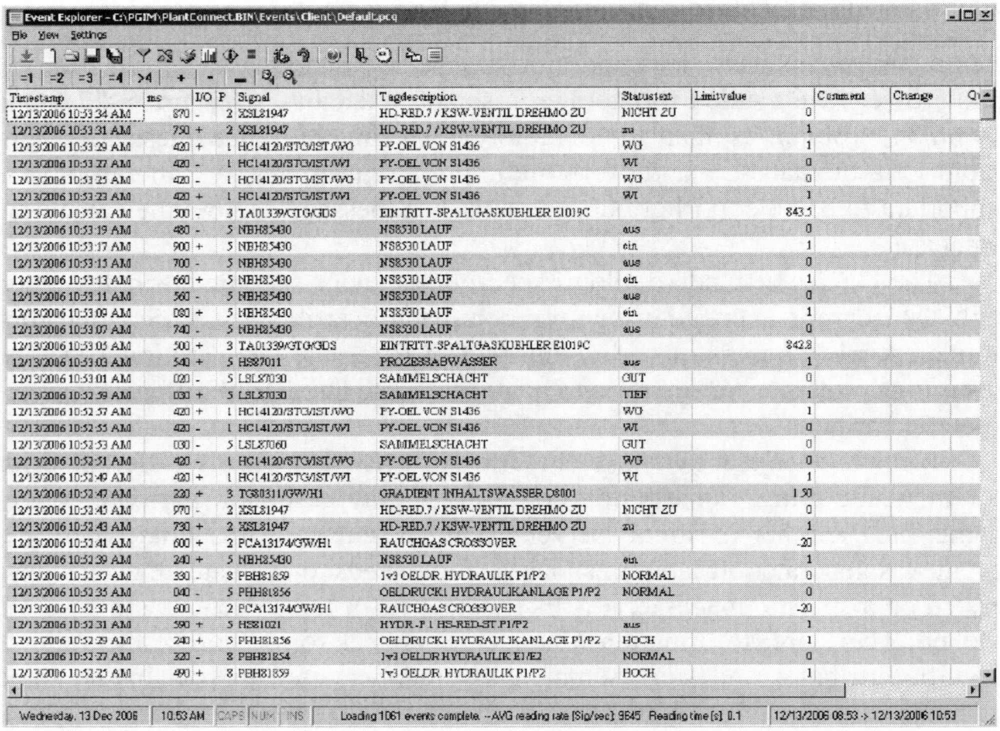

Figure 5–27   Stored Alarms and Events

differences are used to calculate total stress factors. These evaluations are automatically executed on a daily basis.

- Start-up Optimization for power plants. A model of the plant is used for online optimization of start-up procedures. The specific basic process conditions, such as thermal stress of critical thick-walled components and the margins for maximum permissible loads, are predicted and included in a closed control loop. This allows a plant to save 10 to 20 percent of the fuel cost for each start-up (Rode, Franke and Krüger, 2003).

Reporting

Reports are usually based on a prepared report template. Reports can be generated on demand or at cyclical intervals. Reports can be printed on paper or archived as an electronic document. Typical reports include:

5.5 Information Management

- Production status reports for managers
- Compliance reports for regulatory agencies
- Detailed production reports for FDA validation
- Status reports for operations
- Shift hand-over reports that show what has happened during the last shift and can be used to support hand-over discussions

Real-time data, historical values, lab data, batch information, and event information can be incorporated into reports created in Microsoft Excel, Crystal Reports, or another report package that uses ODBC (Open DataBase Connectivity) or Microsoft OLE DB (Object Linking and Embedding, Database) data access.

Information Distribution

One important function of Information Management is to distribute information beyond the control room. Users including the plant management or process engineers want to work from their offices but need to have access to (near) real-time plant data. Also, a corporate headquarters that could even be located on a different continent might want to access current plant data.

For security reasons the networks inside and outside the control room are usually strictly separated with the help of firewalls (see Section 2.7). Information Management acts as an important mediator between the zones inside and outside the control room.

Information Management provides a gateway to the world outside the control room (Figure 5–28). On the other hand, many Information Management results are relevant for control room operators also.

If certain important events like an unplanned shut-down or a major disturbance occur, the concerned people need to be informed, independent of where they currently are. Notifications can be sent as an alarm, email, or SMS (Short Message Service) message.

Wide Area Distribution

For distributing information beyond the control room, two different strategies can be used for the client program:

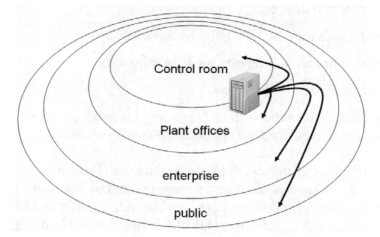

Figure 5–28 Distribution of Information Management Results to Different Zones

- **Thick clients**—These are standard applications that communicate with the Information Management server via TCP/IP. In the past, thick clients could offer richer User Interfaces with faster and better interaction possibilities. However, the fact that they communicate via special ports makes it more and more difficult to run them in a secured environment.

- **Thin clients**—Thin clients that are based on standard Web technology.

Since its invention in 1989, the World Wide Web (WWW) has had extraordinary success. Web browsers are a standard part of everyday life. Web technologies like HTML (Hypertext Markup Language) and HTTP (Hypertext Transfer Protocol) are base technologies used for all kinds of different purposes. Web technology has the following key advantages:

- **Zero or near zero installation**—Installing client software can be a major logistical and administrative effort, especially in larger organizations. A web browser is part of most modern operating systems such as Windows, Mac OS, or Linux. It is very convenient to roll out a new application just by sending a URL to the users. Plug-ins like Adobe Flash™ or Microsoft Silverlight require some installation effort but allow more appealing and interactive applications. The visual and interaction quality of many such web-based applications is as high as of traditional thick client applications. Both packages are part of the standard configuration in many organizations.

5.5 Information Management

- **Firewall-friendliness**—Because of the need to secure more and more segments even inside a company network (see Section 2.7 on security), communication via arbitrary TCP/IP ports is very often blocked by an intermediate firewall. Although it might be possible to open such ports, in many cases this is practically not feasible. A communication via the standard TCP/IP ports 80 and 43 has the highest chance to go through the intermediate firewalls.

- **Automatic updates**—If software installed on widespread clients needs to be updated, this can be a major effort. In case of a web-based solution, only the servers need to be updated, and the URL configured on the clients will automatically point to the updated application.

- **Independence from the operating system**—Web browsers behave the same (or at least very similarly) independent of which operating system they are running on. It is possible to create web applications that even work on browsers on mobile phones or other devices with small screens.

- **Mainstream momentum**—The success of web technology in mainstream computing means that lots of high-quality software modules are available and users are accustomed to and well-trained in web technology.

For these reasons, there is a clear trend toward thin Information Management clients.

Notifications

Some results of the Information Management calculations require the immediate attention of operating or maintenance personnel. For example, if a quality problem has been discovered, the operators should be notified so that they can identify and eliminate the source of the problem as fast as possible.

In some more or less autonomous plants where the operators are informed about problems in the process via SMS, the raw process alarms and events would result in a unfeasibly high number of messages. In such cases the Information Management component can be used to refine alarms & events so that only important messages are sent.

Notifications are typically issued as

- Short message service (SMS) to mobile phones
- Email

- Alarm & event message
- Direct creation of a work order in a Computerized Maintenance Management System (CMMS).

Remote personnel notified of critical events via mobile telephones or e-mail accounts need a means to acknowledge notification and provide confirmation of receipt.

Application Interfaces

Information Management is a platform to acquire, compute, and distribute data. The collected or computed data should be available to applications such as reporting, Spreadsheets, Energy Management or Overall Equipment Effectiveness (see below). The importance of standard interfaces has already been emphasized for data acquisition where OPC plays an important role. OPC is a very good choice not only for data acquisition but also to make Information Management data available for other applications:

- OPC HDA can serve a series of collected or computed historical data from the component. It is a good choice for cascading several components or for data migration.
- OPC DA can serve current results of calculations or data consolidated from various sources.
- OPC AE can serve consolidated alarms and events from various event sources. As the focus of Information Management is on data storage, such a consolidating OPC AE server will usually only contain alarms and events and not offer the full richness of an alarm state machine that includes management of distributed acknowledgments. No historical OPC "HAE" standard has been specified, therefore there is no standard way of accessing historical alarms and events. However, as storing the alarms and events in a database table is straightforward, accessing them via SQL (Structured Query Language) statements is usually quite simple.
- In the future, OPC UA will be an important interface for Information Management, because it unifies OPC DA, AE, and HDA. In addition

OPC UA inherently allows secure wide-area access. OPC UA also provides an infrastructure for modeling plants.

SQL and ODBC are widely used database standards and can be used to access data from the Information Management component.

The interface to powerful spreadsheet programs like Microsoft Excel is part of most Information Management components (see the earlier section on Reporting). Data can be interchanged via drag-and-drop or the spreadsheet may contain a special add-in that allows access to data from the Information Management component.

For automated access via various programming languages, the component needs to offer an Application Programming Interface (API). The example shown in Figure 5–29 shows how an API implemented with Microsoft COM can be accessed from Visual Basic for Applications (VBA™) to read one-hour average values from the Information Management component.

Typical Information Management Applications

Tom DeMarco has coined the phrase "You can't control what you can't measure" (DeMarco, 1982). It not only applies to classical closed loop control scenarios (where it is obvious), but also to higher-level business objectives like quality, effectiveness, and availability. If such values are not continuously assessed and fed back to production as quickly as possible, production usually runs at far from optimal. This is the reason Information Management plays such an important role in a modern CPAS.

Figure 5–30 shows the infrastructure required to connect value-adding applications with process and information sources. Examples for such applications like Energy Management or Overall Equipment Effectiveness are being discussed in the following paragraphs. The manufacturing service bus makes all relevant information available. The layer for industry-specific solutions allows customizing the Information Management system in those cases where general solutions cannot be found. For example, in paper making, cross-profile quality management plays an important role, but has no equivalent in other industries. In the following some of the most typical Information Management applications are described.

```
StartTime = Server.Pct2Gmt(#1/1/08#)
EndTime = Server.Pct2Gmt(#1/2/08#)
Resolution = „1 h"
Compression = „AVG"
SET Werte =
   Server.GetHistIntervalValues(„\\MIW2\West_I\1CGA01\1LAB10
   CF001/VXA", _ StartTime,EndTime, Resolution, Compression)
For Each Wert in Werte
Debug.Print Server.Gmt2Pct(Wert.Time), Wert.Value
Next Wert.
```

Figure 5–29   Example Demonstrating How an Information Management API Can Be Accessed

Key Performance Indicators (KPI)

Key Performance Indicators are quantifiable measurements, agreed to beforehand, that reflect the critical success factors of an organization. One of the most widely used KPI in the context of process automation is overall equipment effectiveness (OEE).

To keep their plants reliable, many manufacturers are setting up optimized maintenance programs to improve their overall equipment effectiveness. The OEE is a success factor for all maintenance and asset management strategies and is therefore an important KPI when it comes to the operation of a plant. It is defined as:

$$OEE = Availability \times Performance \times Quality$$

The gray rectangles in Figure 5–31 show periods of full production. The white rectangles show periods of reduced production, whereas the black rectangles mean downtime. Plant availability can be reduced by equipment failure or by non-operating time for setup and adjustments. The performance of the plant is affected by idling time and minor stops or by reduced production speed or throughput. The quality can deteriorate because of start-up losses and defects in the process, resulting in products that are either waste or have to be sold at a lower price. All three quantities are normalized to 100%. If everything is running at its best point, then the OEE is 100%.

5.5 Information Management

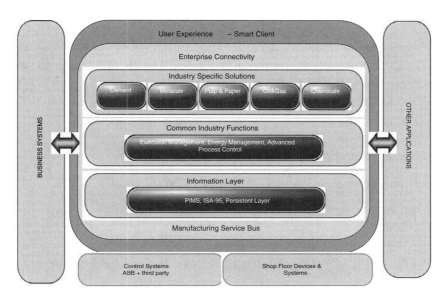

Figure 5–30   Layered Application Architecture

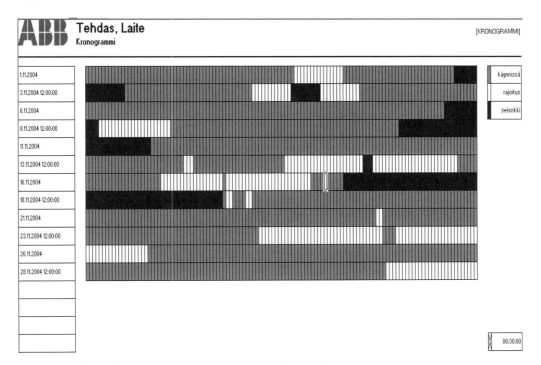

Figure 5–31   Chronogram Showing Plant Availability

Energy Management and Optimization

Energy Management and Optimization applications built on top of Information Management provide advanced tools to support the plant energy consumption from both operational and economic perspectives (see the Section 5.6 for how to connect an Energy Management and Optimization application to an ERP system). Energy management and optimization helps to:

- Forecast electricity, steam, and fuel consumption
- Maximize cost efficiency by load scheduling
- Optimize generated versus purchased electricity
- Manage electricity purchase and sales transactions
- Monitor and control peak loads, energy balance, and efficiency
- Support decision making with simulations and "what-if" analysis capabilities

Figure 5–32 shows the energy consumption of different parts of a paper plant. Consumption forecasts for paper machines are derived from the planned production grade and rate, which may be entered into the system from the production planning system through an interface. Mechanical pulp production is a big energy consumer, and its energy consumption is dependent on the running state of the refiner and grinder units. Some consumers follow weekly load profiles, which can be used in their consumption forecasting. Energy Contract Management selects the energy resources to match the time-varying energy consumption with energy supply at optimal cost. Many plants were built in times with low priced energy, so usually there is lots of potential for improvements. For example, in one Swedish paper plant it was possible to reduce energy consumption for pulp production from 850 kWh/ton to 635 kWh/ton.

Energy management and optimization is a good example where in addition to real-time and historical data, forecast data needs to be handled. Electrical energy cannot be efficiently stored and is therefore much cheaper if consumed exactly when purchased. Real-time monitoring and reporting functions include basic on-line calculations, such as energy balances, efficiencies, and other performance figures, consumption of fuels and chemicals, and emissions. Energy balance and consumption data is also used for cost tracking to allocate energy costs to users.

5.5 Information Management 311

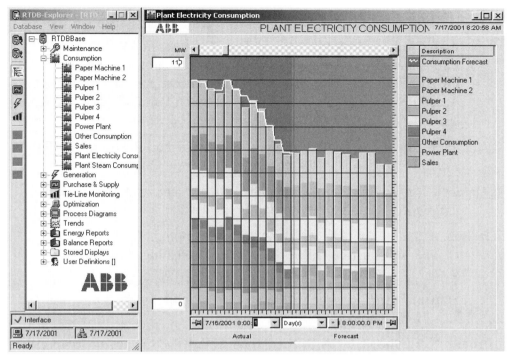

Figure 5-32 Energy Management Visualizing the Energy Consumption of Plant Devices

Data Reconciliation

Process optimization requires an accurate process model and reconciled process data. Data reconciliation adjusts process measurements that contain random errors by having them satisfy material and energy balance constraints, and is a way to improve the quality of the measurements taken from a process. Process measurements are inevitably corrupted by errors during the measurement, processing and transmission of the signal (Narasimhan and Jordache 1999). Data reconciliation applications allow the process site to more accurately analyze the quality of measurements and produce a single consistent set of the most accurate reconciled data possible. Data reconciliation also allows the process site to manage the measurement system from the top down, to fix measurement problems at their source, and to generate a single set of data for all application areas.

References

Alsmeyer, F. "Automatic Adjustment of Data Compression in Process Information Management Systems." In Marquardt, W. and Pantelides C. (ed.) Proceedings of the 16th European Symposium on Computer Aided Process Engineering, 2006. Garmisch-Partenkirchen, Germany available at http://bit.ly/tWouV (last retrieved 9 May 2009).

Bristol, E.H. "Swinging Door Trending: Adaptive Trend Recording?" *ISA National Conference Proceedings,* 1990. New Orleans, LA.

DeMarco, T. *Controlling Software Projects: Management, Measurement & Estimation.* New York: Yourdon Press, 1982.

Narasimhan, S. and Jordache, C. *Data Reconciliation and Gross Error Detection.* Houston: Gulf Professional Publishing, 1999.

Rode, M., Franke, R., and Krüger, M. "Model Predictive Control for Boiler Start-up." *ABB Review* 3/2003; available on the Internet at http://www.abb.com/abbreview.

Thornhill, N. F., Shouk, M. A. A., and Shah, S. L. "The Impact of Compression on Data-driven Process Analyses." *Journal of Process Control* 14 (2004) pp. 389–398.

Watson, M. J., Liakopoulos, A., Brzakovic, D., and Georgakis, C. "Wavelet Techniques in the Compression of Process Data." Proceedings of the American Control Conference, 1995, Seattle, WA.

5.6 Enterprise Connectivity

Margret Bauer and Sascha Stoeter

Introduction

The functionalities described in the previous sections of this chapter—control, monitoring, asset management, and information management—are only some, though important, aspects of managing and automating the operation of a production site.

Enterprise Resource Planning (ERP) systems store information about these resources and also map business processes reflecting how the resources are transferred from one state to the next. For example, a material stock is purchased, used in production, and then sold as a product. The ERP system follows the material through these transitions. It centrally manages data required by different departments and functionalities, for example for sales, logistics, maintenance, and production.

Manufacturing Execution Systems (MES) bridge the gap between the high level managerial information of the ERP and online production information from the process that is managed in the control system. MES covers functionalities such as maintenance, quality, and labor management but also detailed scheduling, performance analysis, and production monitoring. The boundaries between MES and ERP systems are not always clearly defined, and there is often a degree of overlap.

A three-level hierarchy consisting of ERP, MES, and CPAS can be found in most production companies. All system levels support decision-making to enhance production efficiency. The improvement objectives, however, have a different focus at each level. The function of ERP and MES is to manage production, that is, to administer, allocate, and plan. Decisions are made strategically to fulfill demand, satisfy customer needs, minimize cost and, ultimately, to maximize profit.

In contrast, the CPAS aims at achieving stable and safe production. The objective of control is to remove unwanted variability from key process quantities. The CPAS is therefore concerned with data acquisition as well as with monitoring and stabilizing the operation. Besides these different objectives, the time scale on which the systems act and plan varies by at least one order of magnitude. ERP systems information is updated weekly or monthly, MES systems usually daily or hourly and the CPAS within seconds or minutes, sometimes even within a fraction of a second.

Enterprise connectivity aims at integrating the systems on different levels. While ERP and MES are reasonably well integrated, the interoperability between control and higher systems is only now starting to develop (Sauter 2007). The different objectives, management versus control, and different time scales hinder the integration. There are even some who argue that planning & scheduling and control cannot work together (Shobrys and White, 2004) because of the opposing mindsets of the people using the systems.

A Motivating Example

The benefit of integration through enterprise connectivity, however, is significant. To illustrate integration, consider a lot of paint that is produced in a batch process. Information about this lot of paint exists in many systems. First of all, the paint will be produced because a customer order was entered by sales personnel. It specifies quantity, color, physical properties, and probably a due date. In a batch management system, the production personnel store the recipe for producing the paint according to the specifications. The scheduling department receives the order data and generates a detailed schedule in the scheduling system, specifying when the paint will be produced. The operators view and control the measurements during the production of the paint, for example, temperature measurement in the reactor in which the paint is mixed. The quality department monitors the measurements, determines the quality, and compares it to the customer specifications. A tracking system tracks the paint further through shipping or the next production step.

The information generated and handled in the various systems is interconnected, often as input and output. Often, duplicates of the information are created. Having all plant information available everywhere gives a faster response time and ensures data consistency and validity. A faster response time may eliminate idle waiting times and thus has the potential to increase production significantly. Also, if intelligence is added to the integrated systems, actions and tasks that were previously triggered manually can now be triggered automatically. Data consistency is ensured through the introduction of paperless production reporting that prevents potentially illegible hand-recorded production data from being mis-entered into the ERP.

Looking Back in History

The idea of integrating different system levels is not new. In the 1970s, the term computer-integrated manufacturing (CIM) was coined to describe industry efforts to move away from "island" solutions by integrating these existing solutions. However, CIM quite spectacularly failed, as articles such as "200 Years to CIM" (Jaikumar, 1993) reveal. Since then, the integration effort has nevertheless been driven further, supported by more standardized technologies on both the ERP and the CPAS side. With advances in standards, CPAS and ERP having expanded from both sides and now cover more and more MES functionalities. CPAS now communicates via OPC and Ethernet, while the number of ERP providers has diminished, with the remaining players offering configurable connectors.

The foundation for a renewed attempt toward integration began in 2000, when the International Society of Automation (ISA) devised standards for enterprise and MES integration, called the ISA-95 series. The International Electrotechnical Commission (IEC) picked up the standards and will publish them as IEC 62264.

The standards describe models and terminology for determining the information exchange between different system levels. Figure 5–33 shows the system levels as defined in the standards. The terminology applies to nearly all production companies, with different vocabulary for batch, continuous and discrete production. Levels are defined as follows: Level 0 represents the actual physical process, while level 1 systems sense and manipulate, and level 2 systems monitor and control. Level 3 systems manage the work flows, while level 4 systems manage the production organization. In this chapter, parts of the ISA-95 series are included to explain the concept of vertical integration and enterprise connectivity.

Integration

The term enterprise connectivity implies the importance of communication among several coupled systems. As the capability of information systems grows,

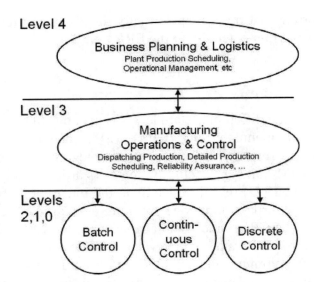

Figure 5–33 System Levels as Described in the ISA-95 Series Functional Hierarchy (*ISA-95 Series, 2000–2007*)

companies become ever more interested in obtaining the most up-to-date status information from all areas to stay current and gain a competitive edge.

For instance, sales people gain improved insights into the committed factory capacity and can negotiate orders knowledgeably. Management can analyze productivity and profitability using the current state of production from the ERP system. Maintenance crews can obtain detailed historic data on equipment and can schedule meaningful forward-looking inspections.

Integration holds several challenges. The systems need to be connected such that information from one system can be made available in another. Furthermore, proper mappings have to be set up to link related bits of information in the systems, where the identifiers and data representations differ. As mentioned before, another challenge in creating the link is translating among different levels of information granularity: from seconds to months and from managing single equipment to managing a complete production site.

Connectivity Structure

The more that hitherto-isolated information systems are equipped with communication interfaces and connected to the ERP, the more important is the structure used for integration. Consider Figure 5–34: It depicts a system with a

5.6 Enterprise Connectivity

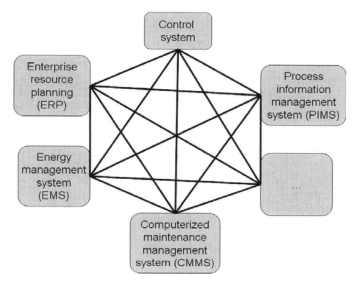

Figure 5–34 Chaotic Connectivity: Point-to-Point Connections Quickly Become Unmanageable

complex connectivity graph despite a small number of endpoints. Setting up further connections only adds to the frustration an engineer would face if asked to troubleshoot a problem.

Point-to-point connections add complexity in a system integration scenario. In order to connect two systems, a dedicated adapter must be developed. On a software level, the adapter translates between the two systems' interfaces and protocols. Often a direct physical link must also be established, unless an existing communication link is available (e.g., LAN—Local Area Network).

With a central point of connectivity as depicted in Figure 5–35, handling a large number of systems becomes manageable. Instead of chaotic interconnections in between any or all systems, all systems connect to a central enterprise connectivity service.

Assuming n systems that all need to be interconnected, the centralized topology reduces the complexity immensely: instead of $n \times (n-1)/2$ connections, only n connections must be established. In practice, not all systems are required to talk to each other, but the complexity of the central approach is never going to be worse.

One place where this complexity reduction provides immediate advantages is in the area of adapters. Fewer adapters are required, and each one connects from a custom system interface to a common interface on the connectivity

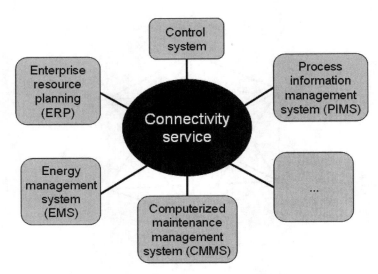

Figure 5–35 Central Connectivity Service Having Exclusive Connections to Each System

service. Instead of mastering multiple interfaces, an adapter developer only needs to be an expert for two interfaces, significantly reducing costs and speeding up integration time. In an ideal scenario, the interfaces would be standards-based to ease the learning curve.

It is true that the centralized approach offers a single point of failure. If the connectivity service fails, the whole enterprise connectivity solution is down. While at first it might seem that a non-star topology could provide robustness through redundant network paths, a closer investigation paints a bleaker picture. The connections between systems cannot necessarily sustain the added traffic, nor is the required routing intelligence built into the systems. There is also the question of how many of the connected systems can fail before the process on which they collaborate must be stopped. Often the answer is: none. Unless all the systems and their connections are redundant, nothing is gained over the centralized approach.

It is also important to note that a failing connectivity service only restores the system to its original state, with multiple isolated systems. If the enterprise can reactivate the old processes, production can continue, albeit in a less efficient way. Information buffers placed at both ends of each connection can help postpone the impact of a failure and give engineers valuable time to solve the problem. Instead of using live data, the disconnected systems operate on cached data. Care must be taken, of course, to synchronize caches with live system data as each connection comes back online.

Integration Standard

When providing a structured connectivity solution as discussed in the previous section, external systems must be mapped to the central connectivity service. This task is simplified if for each connection, the representation at the end of the connectivity service is the same; ideally a standard interface is chosen that is accepted in the IT community. Employing such an interface allows for faster integration and increases the chances of success for the product as more engineers are likely to be familiar with the technology.

The ISA-95 series of standards define objects for entities (e.g., personnel) and concepts (e.g., product segments) found in all types of production. Because no implementation hints are given, the Forum for Automation and Manufacturing Professionals (WBF) crafted the Business To Manufacturing Markup Language (B2MML). It is an interpretation of the ISA-95 series in the form of XML (Extensible Markup Language) schemas. B2MML thus provides a concrete interpretation on which products can be based. As a positive side-effect, the schemas resolve the ISA-95 series' ambiguities. In the following, referring to the standards' modeling elements always means the B2MML interpretation of the ISA-95 series.

Originally intended as a description of the interface between ERP and MES, the ISA-95 series is also well-suited for modeling any system in need of integration. Specifically, the resource models of the ISA-95 series offer the necessary details for the integration of a system. Two uses of the standards must be differentiated in the context of integration: the external system itself and the data sent to it over the connection.

To model the system, its functionality is viewed as a remote procedure call. As such it can be accessed by setting parameters, executing a function call, and waiting for an answer. The system itself can be modeled as equipment. The equipment properties represent both input and output parameters as well as the function call. Parameters are directly mapped to equipment properties, their types taken from B2MML's large set of options. If the required type is not available an encoding scheme is implemented using another type (e.g., string). For the function call, the value of a special property indicates whether a call is made. The remote end of the connection interprets that as a function call and triggers the proper internal operation. This interpretation is part of the adapter.

A web service is an example of an external system that provides data to a production process. Often several operations exists on the web service, each of which needs to be set up with a valid set of arguments. The ISA-95 series allows for the nesting of equipment. This can be conveniently used to mimic the structure of the web service and group operations with their parameters.

Implementation Steps

The implementation of an integration project depends on the application, but there are common steps to be followed. A detailed list of steps using the B2MML implementation of the ISA-95 series of standards can be found in Scholten (2007). Here, the key stages of the implementation phase are highlighted:

1. Identify desired interactions between all systems
2. Create B2MML entities in modeling environment
3. Establish connections to external systems
4. Introduce triggers to model entities
5. Check model for completeness
6. Test solution

In the first step, the relevant information, applications and functionalities are specified. An overview of the information flow is compiled to estimate the complexity of the integration scenario. Entities are created to represent relevant MES objects within the modeling environment. They are created with the help of framework-supplied plug-ins rather than manually to increase modeling efficiency and reduce errors.

For each object, triggers are added that cause a message to be sent from one system to another. This typically involves converting data and translating between the systems' protocols. The trigger can be a Boolean condition over any information available in the MES; in this case the execution of a function is data-driven. The trigger can also be a timer, so that information can be exchanged periodically. The model then needs to be checked for completeness to ensure that all the required information is available. Finally the solution must be tested thoroughly to reduce the risk of malfunction when the MES goes live.

5.6 *Enterprise Connectivity*   321

Here, an example is presented for integrating a web service with an MES using an Enterprise Connectivity framework. In the framework, an MES solution is created by modeling resources and their interactions with the help of a modeling tool as discussed in the previous section. Once the solution is set up, it can be executed. The framework provides dedicated ready-to-use adapters to a variety of external systems, such as various ERPs and databases. Many systems offer a SOAP (originally, Simple Object Access Protocol, now simply SOAP)-based web services interface that enterprise connectivity can address through its generic web services adapter. Each such web service comes with a description expressed through a standardized XML language, the Web Services Description Language (WSDL). It details the service's data types and available operations.

In the following example, a connection to a weather service is established. Having weather information available in the control system is useful for many production environments. For instance, the temperature in open air chemical plants is affected by the outside temperature. Forecasting the weather allows the plant to predict additional heating or cooling requirements. To import the web services functionality into the enterprise connectivity framework, the configuration dialog of the web service adapter is invoked by the user. It only requires a description of the service through a web address or file, and possibly authentication information if requested (see Figure 5–36).

The adapter reads the web service's description and automatically creates entities to represent it in the modeling tool. While it would be possible to create all these entities by hand, the adapter significantly simplifies and speeds up the process of model creation.

The tree structure on the left hand side of Figure 5–37 depicts the new entities. All complex data types are listed under "Equipment Types," while the "Equipment Definition" hierarchically represents the service with all its operations and parameters. Note that the web service is handled here as equipment according to the ISA-95 series and not as a material, product, person, or role, which are the other pre-defined categories.

The right hand side of Figure 5–37 shows the properties of the equipment. In this example, the property of "WeatherImage" is selected. A property can be regarded as the information transferred from one system to another and the characteristics of each property are shown in Figure 5–37 (name, characteristic, type, minimum, maximum, and default value, unit of measure). The checkboxes represent further properties of the equipment. Events, a proprietary extension to the ISA-95 series and indicated by the checkbox "Triggers Event,"

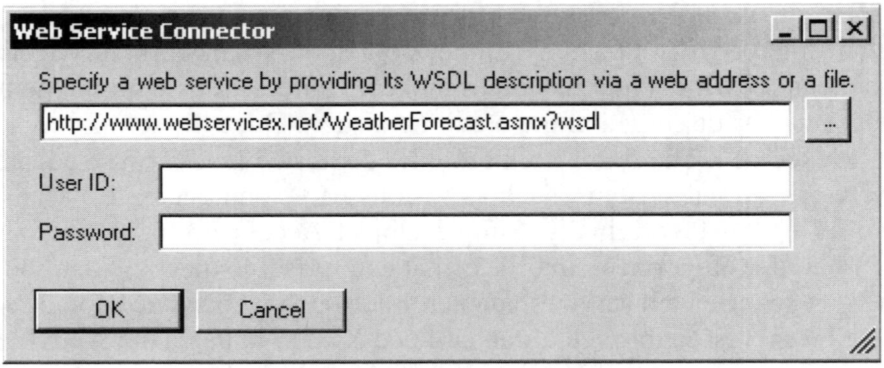

Figure 5–36 Web Services Configuration Screen

allow the adding of custom scripts to the model. For instance, the Invoke event contains code to execute an operation.

The checked Communication property box causes the currently selected "WeatherImage" property to become an OPC property at run-time. This a special feature of the enterprise connectivity framework that is realized through tight integration with the control system.

Figure 5–38 and Figure 5–39 demonstrate this connection. Notice how the structure from the enterprises connectivity is mirrored. Through the "EquipmentProperties" aspect, properties marked as "Communication property" in the framework are accessible and exposed to OPC. This means that any OPC client can now obtain information from business systems and vice versa, allowing for seamless integration of control systems and business systems. The "GetWeatherByPlaceName" operation shown under the EquipmentDefinition in Figure 5–37 is called in the example. The place of interest is set in the Property "PlaceName" to Orlando and the Property "EquipmentEventCall" indicates that this functionality has been executed.

Figure 5–39 shows the result of the call in the ECS Object "Value, ECSObject." The information retrieved from the remote web service is displayed in the bottom right window and shows that the forecast was for a pleasant day on Friday, 18 April 2008, in Orlando. The forecast maximum temperature was 65 and the forecast minimum temperature was 40°F (4.4°C).

5.6 *Enterprise Connectivity* 323

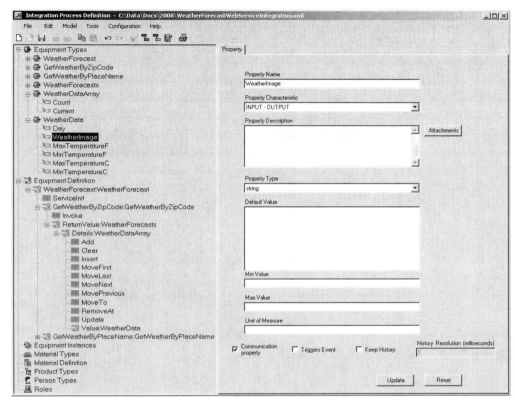

Figure 5–37   Entities Representing the Web Service are Generated Automatically by the Adapter

MES Functionality Through Integration

Designing the connectivity of the various software systems in an enterprise is not an easy task. Every production and manufacturing site is unique and so are the connectivity solutions. There are, however, common tasks and business processes shared by all enterprises as consequences of economic, market, and legal requirements. Production or process control is one of these common functions. The functions are reflected in the business organization, such as in process control, scheduling and quality departments, and in the software solutions used in the company.

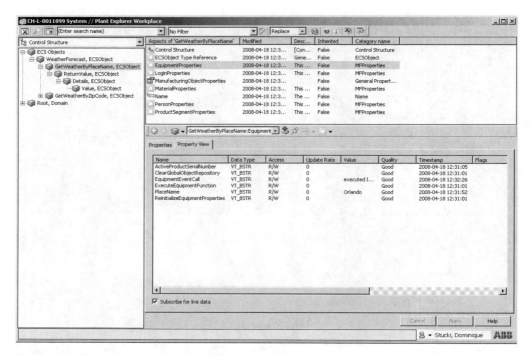

Figure 5–38   Setting up a Web Service Call

The Purdue Model

One of the achievements of the CIM movement in the 1970s is the Purdue reference model that defines ten key functions common to production environments and the interconnection thereof (see Figure 5–40). At the center of the Purdue model is production control; the corresponding system is CPAS. The model therefore has found widespread attention in the control community and consequentially found its way into the ISA-95 series. The shaded area in the Purdue model indicates that the functions in this area are at least partly within the control domain. The arrows pointing from one function to another represent the data flow between these functions. The Purdue model shows the most common, but by no means all, data flows existing between the systems. Integration designers often encounter additional data flows.

At first glance, the Purdue model may appear complex and daunting because of its many functions and data flows. This underlines the earlier statement that enterprise connectivity can be difficult if point-to-point connections are used for data exchange as different systems get involved. To better under-

5.6 Enterprise Connectivity 325

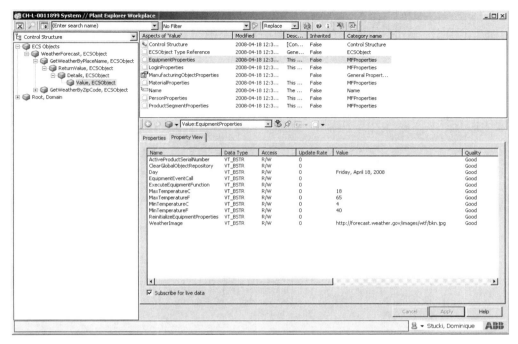

Figure 5–39 Seeing the Results of the Web Service Call

stand the Purdue model, it is helpful to focus on the connection between only two functions at a time. Here, we will discuss briefly some of the specified information exchanges and then in the next section describe the integration of energy management and process control in more detail.

First, consider the data exchange between production scheduling (2.0) and production control (3.0). The scheduling software solution generates a schedule for a daily or weekly time horizon. This schedule is based on the plant's production capability and also on the current production status, including deviations from plan. The schedule is then implemented in production control, for example by retrieving set points from the production schedule. If the connection between the two does not exist, then some manual transfer has to take place. Often, this is done by standard communication tools such as emailing Excel spreadsheets, phone calls, or even by personal meetings.

A second popular example is the integration of maintenance management (10.0) and production control (3.0). All equipment in a plant has to be managed and maintained, as described in Section 5.4 on Plant Asset Management. Computerized maintenance management systems (CMMS) support maintenance management tasks by administrating maintenance work orders,

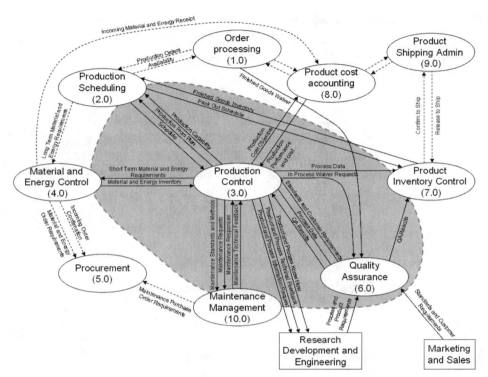

Figure 5–40 The Functional Enterprise-Control Model, Also Referred to as the Purdue Model *(Williams, 1992)*.

as well as maintenance documentation. Some maintenance standards and methods, however, have to be kept in the production control system to be readily accessible by the operating personnel. An exchange of these documentations avoids duplication. Dynamically, maintenance requests stemming from online production control as faults and failures are noticed here first. The requests can be sent automatically to the CMMS and the resulting responses from the CMMS can be sent back to the production control system.

The examples given in the Purdue model are by no means exhaustive. For instance, the data exchange between production scheduling and maintenance would be useful as a production schedule could automatically take into account when a piece of equipment will be maintained and will be unavailable for production. However, it forms a good basis for the development of integration scenarios.

3.2 An Integration Example: Energy Management

For the production of any industrial good, input material and utilities (e.g., electricity, steam, and natural gas) are required. The input material is often procured in bulk, on a monthly or annual basis. The optimal use and planning of energy is crucial for an efficient production facility as energy utilities, electricity in particular, become increasingly expensive. Energy control is largely concerned with electricity load forecasting and detailed load scheduling, that is, how much electricity is required based on a given production plan (see Figure 5–41). The resulting schedule is transmitted to the electricity supplier, either an external company or an on-site facility. It is important that the load schedule is as accurate as possible. Electricity cannot be stored and is therefore cheaper when purchased as predicted. It is ideal to have a predictable load, preferably with a high demand at night when general electricity demand is low.

Industrial production facilities contribute more than one third of the world's energy consumption. In most industries, new or modernized plants typically use less energy than old ones. Energy intensive processes are smelting or heating processes in the metals industry, grinding in mining, concentration and pulping in the paper industry, and distillation and reactions in the chemical industry. Energy management and control is therefore an important task and many companies have their own energy management department or a group allocated to this task in the purchasing department.

Energy management systems are often part of the ERP system but can also be dedicated solutions. The planning has to be carried out on a rolling horizon with an hourly or daily update frequency. Some of the planning is done on an annual basis, especially when contracts with the energy suppliers are concerned. Contracts may contain terms specifying the peak consumption in megawatts that must not be exceeded for a certain time period. It is the task of the energy management system to ensure that the limit is not exceeded.

Figure 5–41 shows a detailed picture of the data flow between energy and production control systems from the Purdue model. The energy control system calculates a load forecast from the production schedule and also based on past energy consumption of the process, in particular deviations from plan due to, for example, unplanned plant shut-downs. The energy control system then generates a detailed load schedule, that is, a detailed forecast of when energy will be used and where. This schedule for the energy inventory is passed on to the production control system and limits are set for the next production. The energy control system can also have what-if scenario capabilities, analyzing

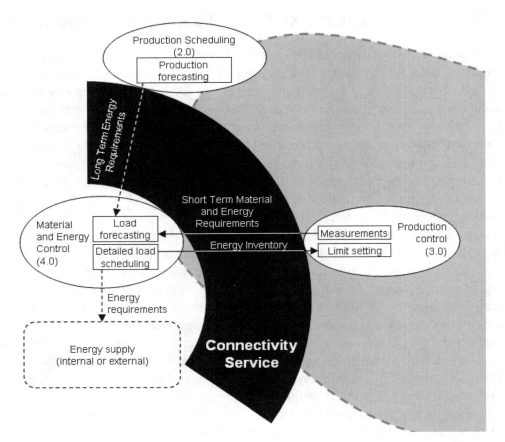

Figure 5–41   Enterprise Connectivity of Energy Management Systems, a Detailed View of the Purdue Model

whether it is better to switch off some part of the production than to exceed an energy threshold. Enterprise connectivity is therefore a vital part of energy control and management.

References

ANSI/ISA-95.00.01-2000, Enterprise-Control System Integration Part 1: Models and Terminology. Research Triangle Park: ISA, 2000.

ANSI/ISA-95.00.02-2001, Enterprise-Control System Integration Part 2: Object Model Attributes. Research Triangle Park: ISA, 2001.

ANSI/ISA-95.00.03-2005, Enterprise-Control System Integration, Part 3: Models of Manufacturing Operations Management. Research Triangle Park: ISA, 2005.

ANSI/ISA-95.00.05-2007, Integration, Part 5: Business-to-Manufacturing Transactions. Research Triangle Park: ISA, 2007.

Jaikumar, R. "200 Years to CIM." *IEEE Spectrum*, 30 (1993) pp. 26–27.

Sauter, T. "The Continuing Evolution of Integration in Manufacturing Automation." *IEEE Industrial Engineering Electronics Magazine* 1 (2007) pp. 10–19.

Scholten, B. *The Road to Integration: A Guide to Applying the ISA-95 Standard in Manufacturing.* Research Triangle Park: ISA, 2007.

Shobrys, D.E. and White, D.C. "Planning, Scheduling and Control Systems: Why Can They Not Work Together." *Computers and Chemical Engineering* 24 (2000) pp. 163–173.

Williams, T.J. *The Purdue Enterprise Reference Model, A Technical Guide for CIM Planning and Implementation* Research Triangle Park: ISA, 1992. ISBN: 1-55617-265-6.

5.7 Planning and Scheduling

Iiro Harjunkoski

Introduction

The planning and scheduling of operations is an area that is becoming more and more critical to many industries, as it provides a systematic method for using the existing facilities more efficiently and economically. This chapter provides a brief overview of planning and scheduling, identifies the main complexities, and describes some of the common methods used to handle them.

Production planning and scheduling play a key role in ensuring the best possible economical operation within an existing facility. Together they deal with deciding upon what to produce (batch, product, intermediate), when (date, time), and where (production facility, machine, production equipment). For many process industry sites, this has been done manually until recently. This means that a scheduler or a team of production planners tries to find a good production strategy for the next days or weeks using pen and paper or an Excel spreadsheet, for instance. This can be a complex task to perform manually due to the large number of possible combinations and frequently occurring equipment requirement conflicts. By using optimization methods of some kind, most commonly mathematical, better schedules can often be generated and the achieved productivity benefits range between 1% and 20%. This may contribute to annual savings or a capacity increase of several million EUR (or their dollar equivalent). The reasons that more "optimal" scheduling has not yet become a standard are manifold:

- Operational margins have been high enough without additional efforts
- Planning and scheduling optimization of industrial-scale processes is not a trivial task
- There are only a few companies that can provide a holistic planning and scheduling optimization concept

5.7 Planning and Scheduling

Furthermore, linking the long-term business targets and short-term operational goals is also far from trivial and requires open systems that ensure the ability to tune and coordinate the functionalities of potentially competing systems, as well as evaluate the achieved benefits on all levels.

Planning vs. Scheduling

The difference between the terms planning and scheduling is not fully clear. Planning can comprise any activities considering how to realize a production; scheduling always has a time aspect in it. Typically planning is used in the context of investments, strategy and other longer-term items. Planning can also be used in all operational considerations where the time concept plays no role. Often, the term short-term planning is used in parallel with scheduling and this is also the main focus in this section, mainly motivated by the close relation to an automation system.

In short-term planning and scheduling, production tasks are allocated to limited resources such as raw material, energy, or equipment. Decisions have to be made concerning the time and resources necessary for the tasks and the order in which the tasks will be carried out. Often, the resources cannot be used in parallel, which makes the time aspect especially important. A problem that illustrates this is taken from the airline industry. An airline possesses a limited number of aircraft that need to be scheduled such that they are at the right airport at the right time and get maintained regularly. At the same time, the non-productive on-ground time should be minimized.

Even more complex is the task of scheduling an airline crew (see Figure 5–42). Here the numerous crew members must be assigned to each of the aircraft. In this problem, lots of factors must be taken into account, such as total working hours, resting times, accommodation costs, possible preferences or combinations of male and female cabin personnel, and returning to the home airport. One can easily understand that timing plays an important role in the decision making as the personnel also need to be at the right location. The allocation task is to assign the crew members to the selected aircraft in order to fulfill the requirements, and correct sequencing makes all this possible and ensures that each crew member starts and ends her/his working period at the desired airport.

Industrial scheduling problems are similar in their problem structure. Instead of crew members, batches, heats or other production entities are considered and

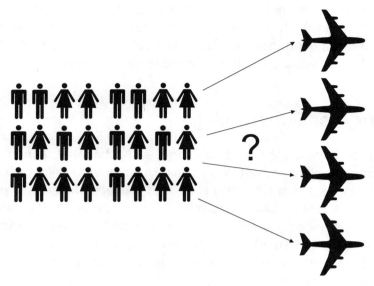

Figure 5–42 Crew Scheduling Problem for Airlines

the airplanes are replaced by equipment, for example a paper machine or a reactor. The constraints are mostly industry-specific and many of them are related to material availability (when), machine set points (e.g., temperature, production speed), and setup and maintenance operations needed for the equipment to ensure efficient and safe production.

In a modern process automation system, scheduling is directly connected to the surrounding environment, such as the control system. A scheduling optimization system may, for instance, collect current status information on the plant through various control system components and after the optimization, provide the actual set points to reach the targeted production plan. This calls for fast access to the relevant data through real-time integration and poses some challenges to the system architecture.

As already mentioned above, the definitions of "planning" and "scheduling" are partly overlapping and there exists some confusion as to which activities should belong to planning and which to scheduling. This question is well motivated since the distinction is not fully standardized and depends heavily on the company, type of production, and how the different business-related tasks are distributed. It can be argued that this distinction can never be standardized due to the number of different standpoints and partially-enforced terminology. Very often, planning and scheduling refer to the same thing: making the logistical decisions needed in order to make the plant produce in a more efficient way.

5.7 Planning and Scheduling

In Maravelias and Sung (2008), planning and scheduling are defined in a useful way as follows:

> *Strategic or long-term planning determines the structure of the supply chain (e.g., facility location). Medium-term or tactical planning is concerned with decisions such as the assignment of production targets to facilities and transportation from facilities to warehouses to distribution centers. Finally, short-term planning is carried out on a daily or weekly basis. At the production level, short-term planning is referred to as scheduling.*

Other useful references are, for instance, Błażwicz et al. (2001) and Conway et al. (2003).

Below, we will focus on the production (scheduling) level and will later make a practical distinction between scheduling and short-term planning. However, before going into a detailed discussion it may be worth spending some time on thinking of how to solve scheduling problems and what makes a good schedule. A production schedule can be derived:

1. Manually
2. Using rule-based methods
3. Heuristically
4. With mathematical optimization
5. By evolutionary optimization algorithms
6. Or by a combination of the above

There are also a number of other approaches, but these are the most-applied ones. The manual method is the traditional one, where planning and scheduling experts use their knowledge to obtain a good production plan. Rule-based methods imply that there are systematic rules that can be followed to reach a good result. This can be also done manually or programmatically using so-called expert systems. Heuristic (or "experience-based techniques") methods make some assumptions, which may not be always 100% true but work for practical problems and help in generating a plan much faster. These and the following methods are implemented with computers.

Mathematical optimization is a theoretically-proven method that can find the best possible schedule for a given plant. The main limitations are the modeling restrictions and possible long solution times. Evolutionary algorithms are a kind of heuristic, and can theoretically reach the optimal solution if it does not have any time restrictions. These methods are mostly used to generate a reasonable (if not fully optimal) solution in a short time. One of the most promising approaches combines some or all of the methods resulting in various hybrid algorithms. In the remaining text, we will focus on mathematical optimization because….

Apart from the method chosen, one also has to have an idea of what a good plan or schedule looks like. The key question is, what is our objective or target, i.e., what do we want to optimize? In mathematical terms, the objective function is a function that expresses the "goodness" of an optimization. This objective function is then either minimized or maximized in the optimization procedure. For example, the total production time, also referred to as the "makespan," or production costs are to be minimized. Alternatively, profit is to be maximized. The objective function is usually more complex and combines several targets, for instance, minimizing the production time simultaneously with energy costs and emissions per unit time. A schedule that finds the best possible solution, that is, the best objective function value, is an optimal one. A close-to-optimal schedule is a solution that lies close to the optimal solution, and a "feasible" schedule only satisfies the production constraints without especially taking into account the optimization targets (see Figure 5–43).

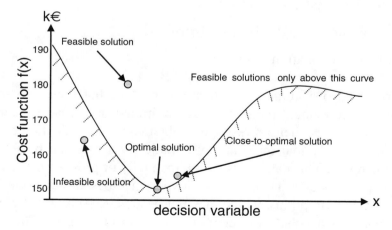

Figure 5–43 Various Solutions to an Optimization Problem

5.7 Planning and Scheduling 335

Some different solution candidates are shown in Figure 5–43. The curve restricts the decision space and all solutions that are valid must lie on or above the curve. Thus, the one below the curve is not valid and is called an infeasible solution.

Scheduling

As we have seen, the most common decisions made in scheduling are:

- Where to produce what, and how much (allocation of resources)
- In which order to produce (sequencing of jobs)
- When exactly to produce (timing of operations)

Practically, this means that the whole matter of production logistics can be seen as a scheduling problem. A scheduling solution must be able to consider (or at minimum not violate) material flows through the plant, take into account all items of production equipment and their current states, and possibly also the use of energy or other limited resources. The problem of equipment allocation is illustrated in Figure 5–44, in which specific machines must be dedicated for each job, possibly resulting in uneven machine loading and the likelihood of reduced total production per unit time.

Scheduling is a critical issue in continuous process operations and is often crucial in improving various aspects of production performance. For a batch process, scheduling also deals with the allocation of a limited set of production resources over time to manufacture one or more products, but the schedule also has to follow a given batch recipe. There have been significant research efforts over the last decades in developing mathematical optimization approaches, but despite significant advances, there are still a number of major challenges and questions that remain unresolved.

In the literature the limitations, strengths, and weaknesses of different mathematical optimization models have been reported. A further issue is often the problem size that one can realistically solve with these models. Some literature reviews can be found in Shah (1998), Kallrath (2002), and Mendéz et al. (2006).

As obvious as it seems, one of the most important aspects of scheduling is time (Figure 5–45). It is crucial to know when an activity starts and ends. The duration can be fixed or also optimized, and it may be linked to production costs or the maintenance needs of an item of equipment. Typically, the relevant

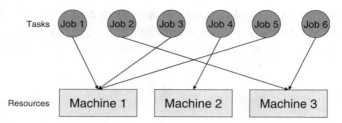

Figure 5–44 Allocation Problem Specifying Which Job on Which Equipment

Figure 5–45 A Block of Time in Scheduling

time granularity or accuracy for short-term scheduling lies within a fraction of a minute to a few minutes, and the scheduling time horizon considered is normally between hours and a couple of days (rarely more than a week). A longer time horizon often makes the problem more complex and difficult to solve. In such a case, the level of details must be reduced. Consequently, the choice of the time horizon is important and largely depends on whether the focus is on short-term planning or scheduling or mid-term or long-term planning, and the details needed.

Scheduling results are typically presented in a Gantt chart (Figure 5–46) displaying the jobs on machines through rectangles in a coordinate system. The horizontal axis normally represents the time and vertical axis the equipment, and the length of a rectangle is directly proportional to the duration of the corresponding job or task.

In Figure 5–46, it can be clearly seen that an item of equipment can only process one job at a time. This is the most common case. Consequently, a job may have to wait for the next available equipment—or a piece of equipment may be idle as it waits for the next job to enter. Scheduling optimization aims at coordinating the equipment in such a way that the overall plant efficiency is as high as possible.

A more industrial style Gantt chart is shown in Figure 5–47. Colors and product/job numbers make the visualization better and the quality of a schedule can be quickly evaluated. In a Gantt view, the equipment load is visible and the identification of a potential production bottleneck may be easy. In Figure 5–47 the bottleneck is most likely the third equipment (AOD), which has no idle times.

5.7 Planning and Scheduling

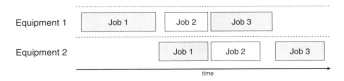

Figure 5–46 A Simple Gantt Chart

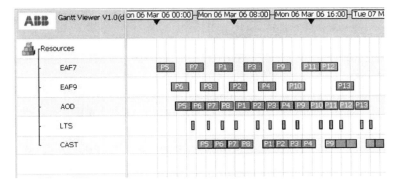

Figure 5–47 Example of an Industrial Gantt Chart

Short-term Planning

In this section we focus mainly on short-term planning that is very close to the process execution. However, short-term planning may also include tasks that do not have the typical scheduling features but are structurally similar and also need to be solved on a similar detail level as a scheduling problem. Such tasks can be, for instance:

- Trim loss or cutting stock problems at paper, polymer and metal industries
- Production grouping problems
- Load planning for delivery
- Short-term energy optimization
- Capacity planning, etc.

Due to the multitude of various planning tasks, a more detailed analysis is not possible within the scope of this chapter. All of the above problems mostly require their own solution approaches and it may be difficult to find a common denominator. The only conclusion is that the above considered planning tasks aim to support the short-term production goals in sub-problems that are not *per se* scheduling problems. These short-term planning tasks are therefore often closely linked to scheduling.

Why is Scheduling Difficult?

Combinatorics (Combinatorial Mathematics)

The main reason for the difficulties in solving a scheduling problem is that the related logical decisions, e.g., assignment and sequencing, are mainly represented by binary (0-1) variables. Consequently, the so-called search space, that is, the number of possible solutions, grows exponentially with the amount of products and equipment. Consider for instance only one machine and three products: a, b and c. These products can be produced on the machine in 3! (three factorial; $3 \times 2 \times 1$) = 6 different sequences (abc, acb, bac, bca, cab, cba). In a normal scheduling problem, we may need to deal with 100 orders and multiple machines. 20 orders alone result in 20! possibilities; it has 18 digits and it is clear that no super computer would ever be able to try out all combinations within a reasonable time. If it was able to test 1 million combinations per second, it would need around 77,000 years to find the best schedule.

The main method discussed in this section is mathematical optimization. The most common problem type is a linear programming (LP) problem. LP problems have linear constraints and a linear objective function. This means that all equations can be represented by lines and the general mathematical form of a linear program is

$$\max c^T x \qquad (5.1)$$

$$Ax \leq b, \quad x \in \Re \qquad (5.2)$$

Above, the coefficients A, b and c are pre-defined parameters and x is a decision variable. The objective function (1) defines a cost coefficient for each variable. The main task is to find those values of x that maximize the objective function (1) such that all constraints of type (2) are satisfied. This can be seen below in a concrete LP problem (P1), where the objective function is maximized subject to three constraints. The two variables are positive and continuous.

$$\max 4x_1 - x_2$$

5.7 Planning and Scheduling

$$7x_1 - 2x_2 \leq 14$$

$$x_2 \leq 3 \quad \text{(P1)}$$

$$2x_1 - 2x_2 \leq 3$$

$$x_1, x_2 > 0$$

A simple LP problem can basically be solved graphically using a pen and paper (as seen in Figure 5–48) but there exists an efficient and robust method, the so-called Simplex method, which is still, after around 60 years of its existence, the most-used mathematical optimization method in the world. But let us first take a look at Figure 5–49. Here, the lines show the constraints in (P1) and the coordinate system axes also bound the positive variables. Thus, the so-called feasible region, where possible solutions lie, is the area restricted by the constraints. The arrow shows the direction in which the objective function is maximized. From mathematical programming theory we know that in an LP problem, the optimal solution is one or more of the corner points. The corner points (circled) are the so-called Simplex points. Here, we can directly identify five candidates. All other solutions between them are feasible solutions. Without working through the mathematics, the optimal solution (2.9, 3) is found in the upper right corner.

If one or both of the variables are discrete variables (0, 1, 2, 3, …) the problem's characteristics change completely. By changing the variable definition

$$x_1, x_2 \in Z^+$$

which means that both variables are now discrete, the problem becomes a mixed integer linear programming (MILP) problem, which is the same as an LP, but has one or more discrete variables (see Figure 5–49).

Now, only the discrete points are accepted as solutions. Most of the earlier Simplex points are thus not feasible solutions and the optimal solution (2,1) is found in a completely different location. Referring to the earlier text, discrete or binary variables are often needed for representing logical decision in scheduling problems and therefore MILP problems are very common in scheduling. The problem becomes a combinatorial problem.

MILP problems are most often solved using the Branch-and-Bound (B&B) method. The B&B method solves a sequence of linear programming problems

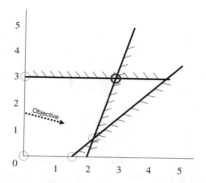

Figure 5–48   A Simple Linear Programming Problem (P1) Illustrated

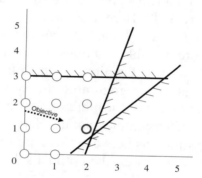

Figure 5–49   Problem P1 as a Discrete Optimization Problem (MILP)

(LPs) and is capable of finding an optimal solution by examining only a fraction of the search tree, as opposed to the brute force method. The B&B algorithm can eliminate whole branches of a search tree (see Figure 5–50) by comparing the potentially best solutions of branches with the already found ones. Thus, instead of having to wait for years, most small scale problems can be solved within seconds or minutes.

In Figure 5–50, a search tree for binary (0-1) variables is shown. The B&B algorithm works also for any kind of discrete variables. In this case, a node can be divided into two parts by adding two new inequalities (above we added equalities) into the subproblems. For example, a value $x = 4.7$ would result into two sub-nodes, where the left node gets the inequality $x \bullet 4$ and the right node $x \bullet 5$. Note that since x is integer, all values between 4 and 5 are infeasible and can be eliminated. Out of curiosity, any discrete variable value can be expressed through a linear combination of binary variables. Therefore binary variables are mostly used in the discrete optimization context. For those

5.7 *Planning and Scheduling*   341

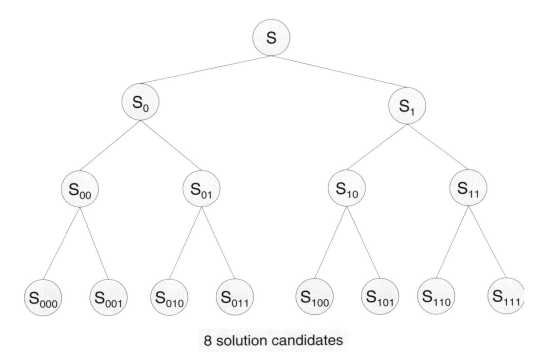

Figure 5–50   A Binary Search Tree

who are interested, more information about solving this type of problems can be found, e.g., in the book by Nemhauser and Wolsey (1999).

Non-linearities

Many scheduling requirements are not by nature linear. Most non-linearities are avoided through the use of logical or binary help variables, for example, in various linear transformations. This is a challenge to the modeling activity, which often ends up in models with a large number of binary or other discrete variables and weak constraints, which makes an MILP problem difficult to solve. To avoid this, supporting heuristics and evolutionary algorithms have been successfully applied. However, the general and main recommendation is always to avoid optimizing anything that does not have to be optimized.

To conclude, the main factors that make industrial scheduling problems very difficult are:

- Large problem sizes for real problems, which often greatly exceed lab-scale

- Complexity of real process constraints, which is also a modeling challenge
- Challenge of embedding the scheduling solution into the plant system landscape
- End-user requirements (e.g., user friendliness and solution speed)

Regardless of these complexities, there are ways to build working planning and scheduling solutions. The next section will present two such examples.

Industrial Planning and Scheduling Problems

In this section, two industrial examples of optimization solutions are discussed. These cases are related to or can be interpreted as classical batch scheduling problems, but as will be seen below, different operational planning and scheduling problems present specific process-related questions that must be solved.

Copper Plant Scheduling

The production of copper is very energy intensive and the process is subject to many disturbances. The main energy need arises in the continuous primary furnace from the melting of the copper ore into matte, which contains around 65% copper and is then further processed in a converter and anode furnace. After the anode furnace, the melt is cast into solid copper anodes. The process flow is shown in Figure 5–51, where the respective copper content per production stage is marked by the grey line labeled "copper concentration" and starting at approximately 28-32%. The process comprises four stages, of which the two stages in the middle have parallel equipment. A batch is the amount of material required to fill one piece of equipment. Every batch must choose only one of the two parallel equipment possibilities as indicated by the arrows. The batch size is physically limited by the smallest equipment and is in fact determined by the optimization. The processing times on a converter depend mainly on the copper concentration of the matte. This may vary by up to 10%, due to the quality deviations on the raw material: copper concentrate. Other factors affecting the processing times and batch sizes are e.g. cooling material and oxygen concentration in the air. Consequently, working scheduling solutions need to be able to simultaneously consider the batch recipes.

5.7 Planning and Scheduling

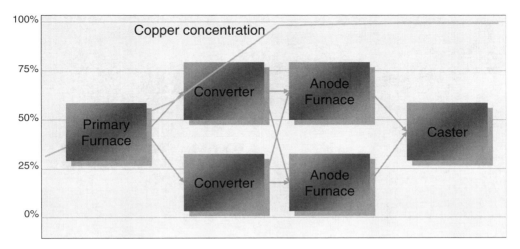

Figure 5–51 The Copper Production Process

The task is to find an optimal schedule for about the next 36 hours. The objective is to maximize the throughput at the caster. The constraints include:

- Avoiding overlapping jobs at equipment
- Allowing enough time between the operations
- Ensuring the correct copper concentration even if the input material quality varies (material balances)
- Other constraints related to the synchronization of the production stages

The main decision variables are the production start and end times, material amounts, and related durations, as well as which batch should be processed on which equipment.

A software solution that fulfills most of these requirements can be seen in Figure 5–52. The information input can be provided manually or, if possible, automatically collected from production databases. Sometimes an automatic approach may not be as attractive due to the fact that the local operators have fewer opportunities to review, evaluate, and modify the suggested schedules. More information about this specific solution can be found in Harjunkoski et al. (2005).

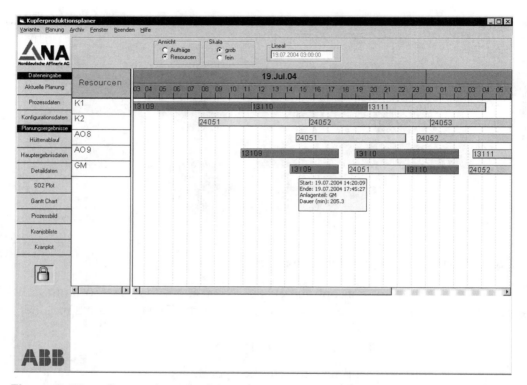

Figure 5–52   Screenshot of a Copper Plant Scheduling Solution

Integrated Re-trimming Module

The other example, quality-based re-trimming for the paper industry, is not a scheduling problem but a short-term planning problem, which can be solved using similar methodologies. The problem is described in more detail in Harjunkoski and Säynevirta (2006). The main idea is to improve an existing cutting plan for a jumbo reel exiting a paper machine. The cutting plan is the outcome of a solution of a stock cutting problem and the objective of the re-trimming optimization problem is to replan the cutting strategy based on additional quality information. This is done during the time that the jumbo reel lines up and waits for a winder. Normally, the reel is divided into sections or trim sets that are cut in width using a derived pattern in order to produce customer paper rolls of the desired length and width. Longitudinal and cross-directional knives allow a cutting pattern to be generated that includes a number of rolls with the same length but varying widths fitting to the jumbo-reel width.

5.7 Planning and Scheduling

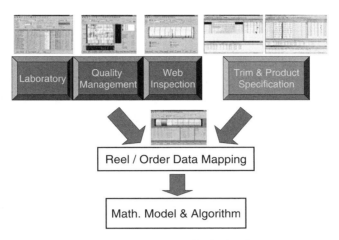

Figure 5–53 System Data for Re-trimming Optimization

If the jumbo reel length now takes the role of the time and a trim set the role of an item of equipment, then some standard scheduling methodologies can be applied. In this specific problem, we perform the re-trimming based on quality data from the quality control systems (see Figure 5–53) and this information needs to be passed to the solution software. Parts of the paper reel that contain quality deviations may need to be rejected or downgraded, resulting in a value loss. The objective of the re-trimming optimization is to maximize the total value of all customer rolls in the jumbo reel. A value or price is calculated *a priori* for each roll at each position on the jumbo reel and the main decision variable is to decide the optimal positions of the rolls. As in the scheduling problem, here the overlapping constraints are also needed, as well as sequencing constraints between the rolls in a pattern.

A re-trimming optimization module[2] was developed which can be directly integrated into an existing collaborative production management (CPM) system. This approach requires well-standardized interfaces to the hosting application and can be easily applied in other cases. As can be seen in Figure 5–54, all the data and/or solution complexity is hidden from a user.

This second example also touches the topic of usability. A planning or scheduling tool can be immensely complex and the number of questions that need to be clarified for simply modeling a problem may be overwhelming. The user who is expected to use a planning or scheduling tool in his or her daily work should be able to do so as a routine. The user has to focus on the main

---

2. Harjunkoski and Säynevirta (2006).

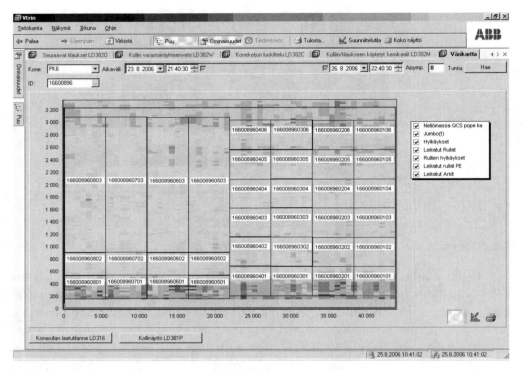

Figure 5–54 Screenshot of a Re-trimming Application

issues of interest to ensure that the production schedule is realistic and combines the theoretical optimization with practical considerations—and, above all, does it within a reasonable amount of time.

Therefore, it remains on the software solution provider to hide the complexity from the user. At the same time, the solution should be as tunable and configurable as possible. The more recognized standards have been applied in the development, the better chance there is to find qualified people who can easily maintain and further develop a running solution. This is extremely important from the product life-cycle point of view, both for the end user and for the vendor.

Potential Benefits

As we have seen, a scheduling optimization solution aims at finding more optimal production strategies. The targets of the optimization are specified by the objective function, underlying logic of the algorithm, and embedded rules

5.7 Planning and Scheduling

that all form the core of the optimization. The objective function is extremely critical to the success of the optimization task, as it alone defines what the "goodness" of a solution is. Very often combined objectives, such as production costs, total production time or makespan, electricity requirements, customer due date violations, emissions, and so on, need to be optimized simultaneously. In these cases, it is important to be able to balance the objectives, for instance by:

- Weighing them against a common basis, for example in monetary terms.

- Replace some of the objectives with hard constraints, i.e., provide predefined limits for some variables and constraints and exclude the corresponding terms from the objective function.

- Methods to reach a Pareto optimality, for instance by changing the objective function components iteratively until an improvement cannot be found.

A successfully configured objective function can help to improve production efficiency significantly. Potential benefits may, for instance, be:

- Lowered production costs (5–30%)

- Shorter total production time (10–50%)

- Reduced emissions (1–20%)

- Lower energy need (5–20%)

- Less delays in customer delivery (10–80%)

The percentages given above provide only a typical range of potential improvement factors. The individual benefits are, of course, case-specific and must be proven on site.

Production planning and scheduling can bring significant benefits to the producer and enable cost savings without any substantial investments other than for a PC and optimization software licenses. This may be vital for existing plants, where revamping is not always economically possible, as well as for green field plants in order to achieve a high-quality planning and scheduling culture from the very beginning.

Integration with a Control System

Integration with a control system has been discussed earlier in this book. Connecting an operational planning or scheduling solution to a control system enables the full benefits, as most of the data can be uploaded automatically, the solution is part of the system infrastructure, and the data used is always up to date. A CPAS should therefore consider the planning and scheduling functionalities as a module or block in its system architecture. Interfaces need to be well defined. It is advisable to follow, as much as possible, widely accepted existing standards, e.g., ISA-95, in order to allow easier connectivity to other systems. The most important issue is to focus on what information the planning and scheduling solutions need and what information the solution should return. As a planning or scheduling activity mostly takes some time, on the order of minutes to sometimes hours, it should also be considered that the important input data fed into the solution should still be valid at the time when the optimization run completes. This is one of the major differences between an offline planning and scheduling solution and an online control system.

References

Błażewicz, J., Ecker, K.H., Pesch, E., Schmidt, G., and Weglarz, J. *Scheduling Computer and Manufacturing Processes*. Berlin; New York: Springer, 2001. ISBN: 3540419314, 9783540419310.

Conway, R.W., Maxwell, W.L., and Miller, L.W. *Theory of Scheduling*. N. Chelmsford: Courier Dover Publications, 2003. ISBN: 0486428176, 9780486428178.

Harjunkoski, I., Beykirch, G., Zuber, M., and Weidemann, H-J. "The Process 'Copper'—Copper Plant Scheduling and Optimization." *ABB Review* (04/2005) pp. 51–54.

Harjunkoski, I. and Säynevirta, S. "The Cutting Edge: Cutting the Inefficiency Out of Paper Re-trimming." *ABB Review* (04/2006) pp. 53–58.

Kallrath, J. "Planning and Scheduling in the Process Industry." *OR Spectrum* 24 (2002) pp. 219–220.

Maravelias, C.T., and Sung, C. *Integration of Production Planning and Scheduling: Overview, Challenges and Opportunities.* Proceedings Foundations of Computer-Aided Process Operations (FOCAPO 2008) pp. 13–22.

Méndez, C.A., Cerdá, J., Grossmann, I.E., Harjunkoski, I., and Fahl, M. "State-of-the-art Review of Optimization Methods for Short-term Scheduling of Batch Processes." *Computers and Chemical Engineering* 30 (2006) pp. 913–946.

Nemhauser, G.L. and Wolsey, L.A. *Integer and Combinatorial Optimization.* Hoboken: John Wiley & Sons, 1999. ISBN: 047182819X, 9780471828198.

Pinedo, M.L. *Scheduling: Theory, Algorithms and Systems.* Berlin; New York: Springer, 2008. ISBN: 0387789340, 9780387789347.

Shah, N. "Single- and Multi-site Planning and Scheduling: Current Status and Future Challenges." *Proceedings of the Third International Conference on Foundations of Computer-aided Process Operations* (1998) pp. 75–90.

CHAPTER 6

6.1 Operator Effectiveness

Tony Atkinson and Martin Hollender

Introduction

It is estimated that the global process industry loses $20 billion, or five percent of its annual production, due to unscheduled downtime and poor quality. ARC estimates that almost 80 percent of these losses are preventable and 40 percent are primarily the result of operator error.

Today most processes are highly automated. Usually a plant can be run with a fraction of the staff level required 30 years ago. But in most cases, visions of totally automated plants that can be operated completely without humans didn't come true. Well-trained and capable staff continues to be a precondition for success. For example, a modern aircraft could be flown 100% automatically, but most passengers would hesitate to fly without a pilot on board. The emergency landing of an aircraft with both turbines broken in the Hudson River in 2009 is an example of how human operators can resolve situations that cannot be handled by automation. In the case of Formula 1 racing, it is clear that only the best overall human-machine system (combination of team, car, and driver) will win the race.

Typical reasons for the importance of human operators include:

- The ability to supervise process and automation and handle situations unforeseen by the designers of the automation system. The creative capabilities of the human operator and the ability to cope with ill-defined or novel situations make human operators irreplaceable.

- The ability to identify upcoming problems as early as possible and resolve them before they can do harm.

- The ability to initiate and organize maintenance activities. Although maintenance is traditionally handled by a separate department, there is

6.1 Operator Effectiveness

an ongoing trend to integrate operation and maintenance activities (see also Section 5.4).

- The capability to handle tasks that cannot (yet) be cost effectively automated. This, however, is a typical engineer's point of view. A system which provides as much automation as possible, leaving only the "leftovers" for the human operator, will quite likely not use the full potential of the human operator.

Human operators in highly complex industrial automation systems, such as pulp and paper mills, power plants, and refineries, continue to play a central role, and it is essential to understand and maximize the collaboration between the control system and the human operator. Adopting a systematic design approach is crucial for reasons of safety and optimum system performance (Pretlove and Skourup, 2007).

In an automation project the investment in the control room workplaces, display and alarm engineering, operator interfaces, screens, etc. is significant and should be assigned similar importance to investments in control logic (algorithms, interlocks, sequences).

Endsley et al. (2003) define the term operational complexity. For example, in modern cars automation has been used to reduce operational complexity for the driver. (New features like sound systems, navigation, communication, and climate control somehow contradict this statement, but in this discussion we focus on the core functionality of driving only). Compared with older cars, where drivers had to understand many of the internal features (e.g., in cars with unsynchronized manual transmission, drivers had to double-clutch), the cognitive requirements for the driver have been reduced significantly. However, the operational complexity for mechanics has increased with overall system complexity. Do-it-yourself repairs, as previously practiced by many car owners, are no longer an option.

Most of today's plants are highly automated and the operator's job is to supervise automation and process (Figure 6–1). There is no single, well-established identifying term for the interface between CPAS and operator. The term Human Computer Interaction/Interface (HCI) is well established in general computing, and a vast body of knowledge exists about it. While much of this can be applied to the interface between operator and CPAS, process control has some specific requirements, and some parts of the interaction might not be with a computer. For example, an operator might call a colleague and ask him or her to close a valve manually. Feedback like noise, vibrations, or vision can

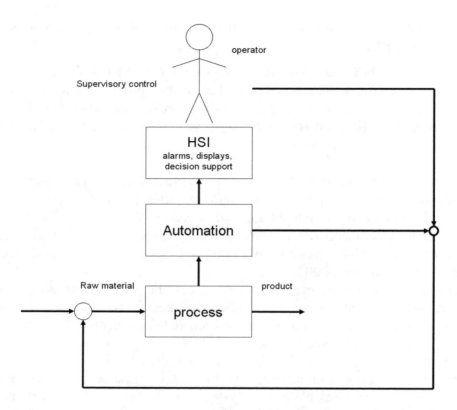

Figure 6–1   Typical Process Controlled by Automation and Supervised by the Human Operator

be a very important basis for decisions, but might not be available from the computer. Many processes have complex inherent dynamics often dictating the (re-)actions of the operators; there is no undo-button for opening a valve.

Other terms that describe the specific interaction between operator, CPAS, and process more precisely are Human-Machine Interface (HMI), Human System Interface (HSI; which will be used in this chapter, because it emphasizes the fact that the interaction can include several humans, machines, computers, and systems), or Man-Machine Interaction/Interface (MMI), but all are less well established.

In the old days of low levels of automation, operators were directly involved with the physics of the plant. For example, when they opened a large valve manually they could directly feel the huge forces involved. In a highly automated plant, operation is much more abstract and in the case of remote operation, the operators are completely detached from the process itself. Under

6.1 Operator Effectiveness

these circumstances, process control might become like a video game. Clicking on a small icon on the screen might have severe consequences in the process. This loss of direct contact with the process needs to be carefully balanced by systematic training.

Some operators even fear interacting with the process, because they know their actions can have huge financial consequences or even impact process safety. An unplanned shutdown of a larger plant can cost hundreds of thousands of dollars. Lisanne Bainbridge (Bainbridge, 1983) has called this the irony of automation: during normal operation the automation does everything and decouples the operator from the process, whereas during process upsets the operator alone needs to understand and manage a highly complex mitigation and recovery process. In the worst case, operator jobs are very boring but very responsible, with little opportunity to acquire or maintain the qualities required to handle the responsibilities. If the automation fails, some operators are not well enough prepared to take over. Simulator training and the switching off of automation layers for training purposes are possible countermeasures.

Many processes are operated around the clock. Operators work in shifts, eight hours or more a day, week by week, year by year (2000 hours per year). This has several important consequences:

- Operators spend a lot of time with the HSI. Design errors will waste lots of performance and can lead to health problems; for example, when operators are continually overburdened.

- Process operation is teamwork. Larger plants require several operators working in parallel. Different shifts need to communicate and maintenance and operations staff need to cooperate with each other.

- After having operated a process for many years, operators are expert users with specific needs. This is one more reason why real operators should be involved in the interface design.

There is often a gap between how interaction was originally planned and the later real day-to-day operation. Examples are average alarm rates that are much too high to handle as discussed in Section 6.2 and nice-looking and colorful displays that are only shown to visitors or management but rarely used for real operation.

It is something of a paradox that the control room operator is often regarded as the "weak link" in the control and safety chain, while simultaneously forming our most flexible resource to monitor and optimize the process and, most importantly, our most sophisticated analysis tool, potentially capable of diagnosing complex and unforeseen process problems affecting the safety of people and the environment.

In many high profile process incidents, the operator had minutes (if not hours) to diagnose and correct problems, but for a number of reasons failed to do so. It is important to recognize that the operator is not necessarily at fault in these cases, and for these events to develop, a number of layers of protection have to have failed. Nevertheless the operator had the opportunity to prevent or mitigate the ultimate consequence of the event and failed to do so.

Analysis of a number of high profile incidents over the previous 25 years of incident recording and analysis reveals a number of failings of the wider operations environment. These can be characterized as:

- **Job Design**—This includes the recognition and documentation of all the tasks that the operator has to perform. Manning levels and competence profiles are directly associated with accurate and appropriate job design. Similarly, the design of the HSI and the control room require that the various tasks that the operator is required to perform are fully documented.

- **Control Room Design**—The design of the control room itself can have a major impact (positive or negative) on the performance of the operations team.

- **Human System Interface**—The HSI is a critical element in delivering the required operator performance. The HSI has to support the operator in all modes of operation (see below) and keep the operator fully informed as to the current operational situation (situational awareness). Loss of situational awareness by the operations team is often associated with the lead-up to a major incident.

- **Alarm Systems**—The alarm system is a key element of not only the HSI but of the process of operating the plant itself. Failure of the alarm system (due to failure to recognize or act upon the alarm, or simply that the alarm fails to annunciate through fault or being suppressed) is implicated in nearly every process event or excursion.

- **Training and Competence**—Clearly the competence of the operator and the operations team has a major impact on the effectiveness of the team in making the timely and correct decisions required for effective operation.

- **Safety Critical Communications**—Communication is a key role for the control room operator. While some communications are associated with efficient operations, others are directly safety related and should be considered with some degree of care and management. The most often quoted example of safety-critical communication is the process of handover from operator to operator at shift change. However, communication up and down the direct chain of management or supervision, between operators in the same control room, or the communication with teams outside the immediate control room should not be neglected.

- **Procedures**—Procedures and their associated support mechanisms (such as checklists) are vital to ensuring consistency of operation and supporting an operator in activities that may be infrequently performed. The best systems of procedure management are used to record learning and support a process of continual improvement.

- **Alertness and Fatigue**—While the alertness and fatigue levels of control room operators are implicated in many high profile incidents, it is not proposed to address this topic in detail here. However, it is worth pointing out a few relevant issues. Fatigue is a cumulative problem that contributes to risk as its effects build up over time. The total fatigue burden on the operator should be taken into account, including activities partially concerned with the workplace (such as commuting) and activities wholly unconcerned with the workplace, but still relevant (such as activities and sleep patterns associated with leisure or family activities). The Energy Institute publishes an excellent introduction to this subject as *Improving Alertness Through Effective Fatigue Management* (Energy Institute, 2006).

Given the nature of this chapter and the context of the book, we will concentrate on the areas directly impacting the control system, specifically the areas of job design, control room design, HSI design and alarm systems. The remaining areas of training and competence, safety-critical communications, procedures, and alertness and fatigue will not be addressed in detail. Nevertheless, they remain crucial elements in effective operations management.

Characteristics of the Human Operator

As human operators don't fit well in the world of mathematical formulas, engineers have a tendency to ignore issues related to human operators. Understanding some characteristics of the human operator helps to design better Human System Interfaces, control rooms, and job tasks. In the following, some important characteristics of the human operator are listed. For a more complete discussion see, for example, Johannsen (1993) or Wickens and Hollands (2000).

Rasmussen (1983) (Figure 6–2) defines three types of behavior or psychological processes present in operator information processing: skill-, rule-, and knowledge-based behavior (SRK). A skill-based behavior represents a type of behavior that requires little or no conscious control to perform or execute an action once an intention is formed. Performance is smooth, automatic, and consists of highly integrated patterns of behavior. In modern plants, most tasks that would require skill-based behavior like manual control have been automated. Most typical operator tasks require behavior on higher levels.

Rule-based behavior is characterized by the use of rules and procedures to select a course of action in a familiar work situation. The rules can be a set of instructions acquired by the operator through experience or given by supervisors and former operators.

Knowledge-based behavior must be employed when the situation is novel and unexpected. Operators need to understand the fundamental principles and laws by which the process is governed. Since operators need to form specific goals based on their current analysis of the system, cognitive workload is typically greater than when using skill- or rule-based behaviors.

Another very important fact is that the capacity of human short-term memory is limited to around 7 ± 2 items (Miller, 1956). This means that an HSI design should not overload the operators' short-term memory, e.g., by forcing them to remember lots of process values. Also, the human capacity to be interrupted by alarms is limited. Modern approaches take this limitation in account and use them to set target values for the CPAS (see Section 6.2 on Alarm Management).

The study of human error is a very active research field, including work related to the limits of human memory and attention and also to cognitive biases. Cognitive biases are strategies that are useful and often correct, but can lead to systematic patterns of error. An example for a cognitive bias is the availability heuristic. It means that many decisions are made on the basis of

6.1 Operator Effectiveness

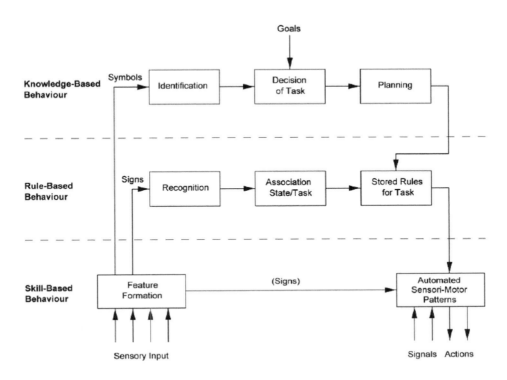

Figure 6–2   Skill-, Rule- and Knowledge-based Behavior *(Rasmussen, 1983)*

information that is readily available rather than on a logical evaluation of all alternatives (out of sight, out of mind). If information is only available on displays currently not shown, it is likely that this information will not be included in the decision-making process. James Reason (1992) has emphasized the fact that not only individual but also systemic and social factors for human error need to be investigated.

CPAS—Goals for the Human Operator

Situation Awareness has been defined (Endsley et al., 2003) as the perception of elements in the environment within a volume of time and space, the comprehension of their meaning, and the projection of their status in the near future. It is a key goal of the CPAS to help the operator to maintain Situation Awareness. For example, large screen displays that show key process information always at the same physical location (sometimes called SDCV: spatially dedicated and continuously visible) are one way to support Situation Awareness.

Many processes run smoothly for long periods of time without requiring any intervention by the operator. It is not easy to keep up sufficient vigilance under such conditions. Sometimes artificial tasks like periodically filling out forms are introduced to keep the operators vigilant.

It might be that some easy-to-overlook symptoms indicate an upcoming problem. If the operators detect these symptoms early enough, a costly event like an unplanned shut-down might be avoided. One important principle in maintaining a high level of vigilance is to avoid false alarms (see Section 6.2). If operators trust their alarm system, an activated alarm is more likely to cause action.

Plants are run by teams. Operational and maintenance staff need to work closely together. Shifts need to be handed over to each other. A CPAS needs to support this kind of teamwork and workflows.

Safety can be enhanced by adding several safety layers (IEC 61511-1, 2003) (Figure 6–3). One of those safety layers is the human operator and his or her ability to drive the process back into normal state. No single layer can be perfect, but the combination of several layers should result in a sufficient safety level.

If developing problems can be detected and eliminated early by the operator, protection layers like automatic shut-down systems don't need to be invoked. If shutdowns and start-ups can be avoided this results in additional safety as those phases are usually critical. It is obvious that the avoidance of unplanned shut-downs not only increases safety but also has high economic benefits.

Modes of Operation

The process operator and associated control system and control room environment typically function in one of three modes depending on the circumstances of operation. These modes of operation can be characterized as normal, planned abnormal, and unplanned abnormal.

Normal operation is associated with routine operations. The role of the operator and the control system is monitoring and optimization. In many ways the operation of the plant in this mode is "automatic" for the operator and requires very little in the way of cognitive resources. It is not necessary for the operator to understand the full implications of the various interactions that are

6.1 Operator Effectiveness 361

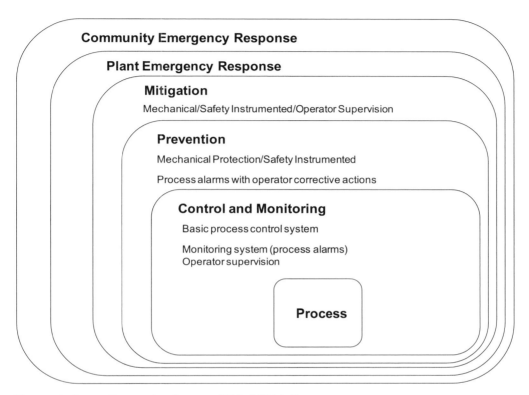

Figure 6–3   Protection Layers *(IEC 61511-1)*

potentially in play at any one time as any necessary corrective control is likely to occur at a relatively simple "single loop" level. In a skill-, rule-, and knowledge-based (SRK) framework, this level of operation corresponds with skill-based behavior. The operator is typically supported by basic training, operational targets, and generic standard operational manuals.

Planned abnormal operation is associated with operations that are not routine, but are foreseen and planned. Operations falling into this category include unit start-up and shut-down, planned maintenance, and abnormal operating modes (such as manual rather than automatic operation of a unit or control loop). Some batch operations, particularly manual interventions not wholly automated or sequenced, can also be thought of as operating in this mode. Important characteristics of this mode of operation include the fact that the mode is foreseen by the designers of the process or operation and that the activity is planned. The role of the operator is management, intervention, and often a high level of communication (for example to personnel outside the immediate control room environment).

Within the SRK framework, this mode of operation is dominated by rule-based behaviors. More sophisticated insights into the interactions of the equipment and process are required of the operator. These should be supported by checklists and procedures, which reduce the need for direct experience by allowing the experience and knowledge of other personnel (usually more senior or qualified operators) to be brought to the task.

Unplanned abnormal operation is associated with unforeseen and/or unplanned deviations from normal or planned operations. Because the process or plant is in a mode unforeseen by the designers of the process, there may be limited or no support available to the operator. The role of the operator becomes diagnosis and management of the situation. The operator requires considerable knowledge of the fundamental principles of the process, equipment constraints, hazardous conditions, and protective systems, and a sophisticated understanding of potential interactions. Within the SRK framework, this mode of operation requires the operator to exhibit knowledge-based behaviors, often unsupported by procedure or even previous experience. Correction of the disturbance may require the operator to reduce throughput or ultimately shut down part or all of the process, so it is important that the operator has the authority and confidence to make this decision where necessary.

Large plants or processes may be in one or more of these modes simultaneously. One unit may be operating normally, perhaps another is down for maintenance, while another is in a disturbed mode. Operators and operations teams must cope with a complex series of scenarios for which a variety of support mechanisms may, or may not, be available.

Job Design

It is common for the control room operator to have a complex job, extending beyond the obvious monitoring and control of the process. It is equally common for the control room environment, the HSI, and the training given to the operator to extend only to operating the process in a normal mode.

Activities that the control room operator commonly undertakes (in addition to operating the process in normal mode) commonly include:

- Startup and shutdown of units or equipment
- Diagnosis and management of faults or process upsets

6.1 Operator Effectiveness 363

- Optimization of the process
- Issuing and managing Permits to Work
- Managing access to the plant or control room
- Answering the telephone (out of hours, all incoming calls may be routed to the control room)
- Coordinating maintenance tasks
- Emergency response coordination
- Handover and communication
- Recording/logging of significant events and deviations

Formal assessment, design, and documentation of the various activities that the operator may be required to perform constitute the necessary first step in ensuring that the operator is best equipped to perform those activities. Techniques such as Hierarchical Task Analysis (HTA) can help understand and document the totality of the operator's role.

Control Room Design

The design of the control room itself will have a dramatic affect on the performance of the operator and the wider operations team. ISO 11064 (1999) contains specific advice on the design process as well as the detailed design and, rightly, emphasizes the need for a structured project process with correctly specified requirements and the widest possible view of the function of the control room.

The job analysis phase will indicate the broad tasks that the operator and operations team will be expected to perform. ISO 11064 defines the control room in terms of "task zones" in which the various activities are carried out. In addition to specific operations-based tasks (e.g., control of the process, issuing of permits, emergency response coordination) consideration needs to be given to "domestic" activities, mess facilities, toilet and washing facilities, changing and locker space, etc.

The control room itself will be the focus of the design process. As with all design processes where a human is to be present in the environment, the design of the control room represents a compromise between conflicting requirements. For example, a linear control desk design is excellent for the operations team members sharing equipment and interfaces but is poor for direct eye contact between operators. The number of operators, the type of tasks to be undertaken, the hierarchy of management and supervision; all will have an impact on the control room layout. Issues such as access control, lighting, HVAC (heating, ventilating, and air conditioning), and the auditory environment will also have an impact on the effectiveness of the control room team, as well as their overall health and well-being.

Control room design needs to incorporate sufficient space for paperwork, procedures, and instructions and other items requiring "desk space," even in a "paperless control room" concept.

Finally, the selection of the non-fixed furniture items will have an important impact on the performance, health, and well-being of the operators. Items such as chairs are subject to a significantly higher rate of use than in a normal office environment. An office chair will be adjusted only occasionally, possible only once by its primary user. A control room chair will be adjusted every shift and, of course, used 24 hours a day. Careful selection of specialist furniture will prevent problems or early failure.

HSI Design

The design of the HSI is an important contribution to the effectiveness of the operator in making timely and correct decisions. The operator typically has many diverse interfaces, even in a modern integrated control room. A typical operator may be coping with one or more screen-based systems, telephones, radio or public-address systems, security and access systems, etc. Given that this book deals primarily with the design of the control system, we will concentrate on that area. However, the total operator burden from all relevant interfaces should always be considered when designing, managing, assessing, or operating a control room.

Most modern operator interfaces follow the Microsoft Windows model, usually running under a version of the Windows operating system. The operator will require a keyboard and mouse (or similar pointer). It is good practice to supply a single keyboard and pointer that can be focused on a number of screens arranged in a "cluster." In newer systems, the combination of screens may eliminate the screen border or interface to allow graphics or windows to span more than one "screen" without a noticeable discontinuity at the "join" (Figure 6–4). This is particularly effective on projection based systems. Where this is avail-

6.1 Operator Effectiveness 365

Figure 6–4 Modern Operator Workstation Demonstrating Seamless Integration of Screens

able, the screen can be thought of as an area of contiguous screen "real estate" where the operator can position windows to meet the immediate need.

The number of screens (or total "real estate") should be chosen to take into account all relevant tasks that the operator is expected to perform. The output of the formal job design exercise will supply the necessary information to allow this decision to be made. For instance, if the operator will be required to monitor the startup of two separate units (with specific task-based graphics) while overseeing normal operation of the rest of the plant (possibly requiring two graphics screens) then the operator interface should consist of a minimum of four screens.

Screen clustering layouts can be single or dual depth. The choice of layout can impact the size of the overall workstation and potential lines of sight within the control room. EEMUA 201 (2002) and ISO 11064 (1999) contain guidance on the design of operator workstations.

There are numerous theories and standards written for the design and implementation of the control system HSI, however, these all represent a compromise in one form or another.

Most control system manufacturers have an integrated approach to at least the "look and feel" of the HSI, often (as mentioned) based around a combination of Microsoft Windows behaviors and a series of graphical standards and design codes based on international standards and compatibility with previous control systems from that manufacturer. These standards are often embodied in graphical "libraries" which can substantially reduce configuration time and therefore overall system cost. This library based approach can also provide a benefit throughout the system life cycle, ensuring that modifications and minor projects are delivered in a fashion consistent with the original implementation.

One overwhelming requirement of the operator interface is the requirement for consistency in the presentation of data. The color, shape, and dynamic behavior of the interface should be defined, documented, and consistent throughout. Consistency of data presentation not only allows operators to assimilate data faster and more accurately, but also helps them to extrapolate from well-known data and circumstances in response to data presented only rarely.

The overall choice of color palette will also represent a compromise between easy-to-use, low-stress color palettes (so-called "cool" graphics) and high contrast (typically black background) color palettes that allow better color discrimination for color-blind operators. Similarly, the density of data displayed on the screen will represent a compromise between the needs of the operator to monitor the maximum quantity of plant information (or put another way, for the HSI to hide the minimum amount from the operator) and the ability of the operator to comfortably and unambiguously read high densities of alphanumeric or graphical information. The needs of the older operator with potentially deteriorating eyesight should also be taken into consideration. ISO 11064 contains advice on the recommended character size based on visual angle.

Where the control system vendor utilizes a specific design or standard for presentation of data, this should be followed.

The screen-based HSI has been a standard for over 30 years and is likely to remain so for some time. There are several features of a screen-based interface that should be understood by both designers and operators.

First, unlike the panel-based operator interfaces that they replaced, the screen-based HSI hides most of the process from the operator at any one time. The screen forms a "window" that can be overlaid onto the process by the operator by navigating from view to view, but this inherently limits the awareness of the operator. This so-called keyhole effect makes it difficult to maintain an overview of the whole process.

Second, the design of the operator interface influences the ability of the operator to perform in the three modes of operation discussed above. The best interface designs enhance operator performance; the worst designs can significantly impede operator performance.

The desired characteristics of the operator interface will vary according to the mode of operation. In normal operation the operator will be monitoring a high number of variables, some or all of which are controlled. Typical concerns

6.1 Operator Effectiveness

are with optimizing efficiency and quality of product and ensuring that there are no significant deviations from target or ideal operating conditions. The best operators will be actively "patrolling" their interfaces rather than relying on the alarm system to alert them to potential problems. In this mode, it is possible for the operator to monitor a high number of variables effectively. Data presentation should highlight deviation from normal operation and ideally give the operator advanced warning of issues and developing problems. Techniques such as trend displays showing "tramline" (high and low operating limits) warning levels, or bulk analog data with deviations highlighted with color should be emphasized.

In planned abnormal operation, the operator is monitoring, controlling, or communicating discontinuous operations, preferably according to a predefined Standard Operating Procedure. The HSI should provide the operator with all the necessary information to perform the task without the need to be continuously changing screen view. This allows the minimum of screen area to be dedicated to the task, allowing the normal operation activities to be performed simultaneously without losing the view of either the normal or the planned abnormal task. It may be that specific "task-based" graphics are required; they should closely align with the associated procedure governing the task. Note that process interlocks and other constraints are often a major feature of this type of operation and where possible the status of these interlocks needs to be incorporated into the task-based graphics to aid in problem diagnosis.

Abnormal operation may last a considerable period of time, potentially spanning a shift change boundary. The HSI should ideally aid an operator in quickly assessing where in the operation the process is.

Techniques such as screen-based checklists (aligning with the relevant procedure), unit status displays, and detailed device status displays (which may be hidden in normal operation) should be emphasized.

As was mentioned, it is common that the control room operator is also monitoring the remainder of the process while supervising or performing planned abnormal operations. The task should be designed so as to not overload the operator or operations team. Sufficient screens and other resources should also be supplied to allow the total number of planned tasks to be performed simultaneously.

In unplanned abnormal operation, the operator is faced with circumstances that may be unique, unforeseen, and outside of his or her experience. Because the situation is unforeseen, this mode of operation is the most difficult to design and plan for.

The role of the HSI under these circumstances is to allow the operator the best chance of diagnosing what may be a very complex and involved problem, giving him or her the maximum awareness of the situation, and ensuring that all possible information is available for consideration. The operations team will, in all probability, be cooperating on the problem, so the HSI needs to support communication and collaboration. Large screens in view of the entire operations team can facilitate the necessary common view to aid team-based communication.

It is a common complaint that in unplanned abnormal operation the HSI affords endless amounts of detail, but fails to deliver the "big picture." Simple overviews of the process dynamics should be provided, such as unit level mass balances, showing inventory in each significant unit and process flows between them. It is important that all potential flow paths are shown on the graphics and that they are not over-simplified as this may lead to some potential scenarios being ignored by the operations team. Trend plots can aid the process of diagnosis when used in conjunction with overview-type screens.

It may also be useful to utilize alternate views of the equipment or process flows. For example, where a heat transfer system is used to supply the heating and cooling services to the process, it is often integrated into the P&ID (Piping and Instrumentation Diagram) drawings and is therefore often integrated into the process graphics. When diagnosing process problems, it may be useful to have a view of the entire heat transfer system unencumbered by the details of the process.

As the process moves through the unplanned abnormal mode, it may enter an automated shutdown mode of operation where one or more units is tripped by the independent Safety Integrity System (SIS). Once this occurs, the role of the operator may change from diagnosis to ensuring that the process has been placed in the necessary safe state by the SIS. Simple status diagrams indicating the desired safe state with comparison to the actual process state can quickly highlight deviations requiring manual intervention.

Task Analysis

Standards like IEC 11064 and the EEMUA 201 guidelines emphasize the importance of a good upfront operator task analysis. Task analysis (e.g., see Kirwan and Ainsworth, 1992) is the analysis of how a task is accomplished, including a detailed description of both manual and mental activities and task and element durations, task frequency, and any other unique factors involved in performing a given task. All too often a task analysis is forgotten during the requirements analysis for a new control room.

Ecological Interface Design (EID)

Ecological Interface Design (EID) is a relatively new approach to designing interfaces for large and complex systems, such as power plants and medical equipment (Burns and Hajdukiewicz, 2004). The term EID was coined in a paper by Rasmussen and Vicente (1989). An ecological interface has been designed to reflect the constraints of the work environment in a way that is perceptually available to the people who use it. By reducing mental workload and supporting knowledge-based reasoning, EID aims to improve user performance and overall system reliability for both anticipated and unanticipated events in a complex system. EID has gained a degree of popularity, particularly where there is a high probability of operation in the unplanned abnormal mode where the operator is required to apply a high degree of knowledge and diagnostic behaviors. However, it is not practical to reserve the techniques of EID for this mode of operation, and the entire interface should be designed according to the techniques and principles involved in EID. However, this may conflict with the control system vendor's design principles or standard libraries.

Operator Participation in the Design of the HSI

As has been mentioned, ISO 11064-1 recommends operator participation in the control room design process, to include the HSI. It is a structured approach in which future users are involved in the design. User participation throughout the design process is essential to optimize long-term human-system interaction by instilling a sense of ownership in the design. In addition, experienced users can offer valuable empirical contributions to HSI design. Their practical experience is not always documented or well known by designers.

Usability Engineering

For the development of software applications, a discipline called usability engineering has been established during the last 20 years. Usability engineering helps to create user-friendly interfaces that allow users to efficiently accomplish their tasks and that users rate positively on opinion scales. Many of its guiding principles can be found in ISO 9241:

- Prototyping
- User involvement

- Expert reviews
- Testing

Usability testing tries to get feedback from real users as early as possible. The subjects are asked to solve real-world problems with the actual or a prototype interface. It is usually easier and cheaper to make changes in the early design phases. If usability problems are found very late it is sometimes not feasible to correct them. Measurable goals help to focus the HSI design team on usability. Examples for such goals could be:

- Average less than one alarm during 10 minutes in normal operation
- Software package can be installed and configured by a typical project engineer in less than 30 minutes
- Operators rate HSI with at least 4 on a 5-point user satisfaction scale

Human System Interface Guidelines

As has been suggested earlier in the chapter, guidelines and standards capture best practice and help to design better Human System Interfaces. For Human System Interfaces in the process industries, four different layers of guidelines exist:

1. General guidelines are based on the characteristics of human cognition and apply to all kinds of Human System Interfaces including industrial control rooms. Examples of such guidelines include ISO 9241 and the Microsoft Official Guidelines for User Interface Developers and Designers (Microsoft, 1999).

2. Industry-specific guidelines like EEMUA 201 or ISO 11064 focus on the specifics of industrial process control.

3. Many production companies have their own internal style guides that define issues like background colors, logos, and fonts for the Operator Stations used in their plants.

4. In some cases individual plants or sites create specific guidelines.

The alarm system is an important subset of the Human System Interface in process control and will be discussed in Section 6.2.

Window System Guidelines

As most operator stations now run on Microsoft Windows operating systems, the general guidelines published by Microsoft (Microsoft, 1999) should be applied whenever it makes sense. They consist of the following four parts:

- Fundamentals of Designing for User Interaction
- Windows Interface Components
- Design Specifications and Guidelines
- Appendices and References

Many operators run computers at home and expect consistency with what they are used to in their work environment. Microsoft had much broader markets (e.g., office and home users) in mind when writing the guidelines so that not everything fits 100% to the very special situation in process control, which represents only a very small market fraction for Microsoft. In the future it is likely that new standards evolving from other general-purpose applications (e.g., the web or smart phones) will be absorbed by CPAS.

ISO 9241

The standard ISO 9241 was originally called "Ergonomic requirements for office work with visual display terminals" and has been renamed to "Ergonomics of Human System Interaction." It is valid for all kinds of application areas, not just CPAS. ISO 9241 consists of the following parts:

Part 1: General introduction

Part 2: Guidance on task requirements

Part 4: Keyboard requirements

Part 5: Workstation layout and postural requirements

Part 6: Guidance on the work environment

Part 9: Requirements for non-keyboard input devices

Part 11: Guidance on usability

Part 12: Presentation of information

Part 13: User guidance

Part 14: Menu dialogues

Part 15: Command dialogues

Part 16: Direct manipulation dialogues

Part 17: Form filling dialogues

Part 20: Accessibility guidelines for ICT equipment and services

Part 110: Dialogue principles

Part 151: Guidance on World Wide Web user interfaces

Part 171: Guidance on software accessibility

Part 300: Introduction to electronic visual display requirements

Part 302: Terminology for electronic visual displays

Part 303: Requirements for electronic visual displays

Part 304: User performance test methods for electronic visual displays

Part 305: Optical laboratory test methods for electronic visual displays

Part 306: Field assessment methods for electronic visual displays

Part 307: Analysis and compliance test methods for electronic visual displays

Part 308: Surface-conduction electron-emitter displays (SED)

Part 309: Organic light-emitting diode (OLED) displays

Part 400: Principles and requirements for physical input devices

Part 410: Design criteria for physical input devices

As an example of how ISO 9241 may be applied, Part 12 deals with the use of colors. It is recommended that color should never be the only means of coding because some people discriminate certain colors poorly or cannot discriminate on the basis of color at all. Color is a good auxiliary code. It should be made redundant with some other coding techniques.

6.1   Operator Effectiveness   373

The important Part 110 "Dialogue principles" defines the following principles:

- Suitability for the task
- Self-descriptiveness
- Conformity with user expectations
- Suitability for learning
- Controllability
- Error tolerance
- Suitability for individualization

ISO 11064

The ISO Standard 11064 "Ergonomic design of control centers" consists of the following seven parts:

   Part 1: Principles for the design of control centers

   Part 2: Principles for the arrangement of control suites

   Part 3: Control room layout

   Part 4: Layout and dimensions of workstations

   Part 5: Displays and controls

   Part 6: Environmental requirements for control centers

   Part 7: Principles for the evaluation of control centers

Part 1 defines the following design principles:

- Application of a human-centered design approach
- Integrate ergonomics in engineering practice
- Improve design through iteration
- Conduct situational analysis

- Conduct task analysis
- Design error-tolerant systems
- Ensure user participation
- Form an interdisciplinary design team
- Document ergonomic design basis

EEMUA 201

The EEMUA Guideline 201 "Process plant control desks utilizing human-computer interfaces" was published in 2002. The scope of EEMUA 201 is:

- The factors to take into account when designing an HCI for industrial processes
- Display hierarchies
- Screen display format design
- Control room design

Key design principles defined in EEMUA 201 are:

- The primary function of the HCI is to present the operator with a consistent, easy-to-use interface that provides monitoring and control functionality under all plant conditions.
- A task analysis should be performed to capture the full scope of the operator's role.
- Mandate complete end-user involvement in both detailed display format design and the overall system design.
- The number of physical screens provided in the HCI must allow for complete access to all necessary information and controls under all operational circumstances.
- The HCI design should allow for a permanently viewable plant overview display format.

- Continuous access to alarm indication should be provided.
- The capability to expand the number of physical screens should be built into the original design.

Conclusion

In conclusion, it may seem that the operator is partly responsible for, or is implicated in, almost every process industry event. While this is undoubtedly true on the superficial level, two things should be remembered. First, the operator is the product of many elements, not least the support he or she is given by training, experience, supervision, procedures, and other support mechanisms, the HSI and alarm system, and the very culture and environment he or she works in. In the authors' experience, operations personnel are conscientious and where errors are made it is often in pursuit of maintaining production or quality.

Second, it is important to realize that while the process industry, its associated regulators, and even the media are very good at recording the rare instances where things go dramatically wrong, we are collectively much poorer at recording where things go right. For every accident or event, there are untold circumstances where a control room operator makes the correct and timely decision that saves lives, maintains the environment, or contributes to the business bottom line.

Human operators have an important role in the supervision and control of highly automated, complex processes. Rising automation levels, use of computer technology, and staff reductions have made the job more abstract and have decoupled operators more from the process. In case of unforeseen events, the creativity and flexibility of the human operator are required to get the process under control again. Systematic training and high-quality Human System Interfaces are needed to support the operator during such a challenge. Knowledge about human characteristics and the related standards and guidelines helps designers to create high-quality Human System Interfaces and control rooms.

References

Bainbridge, L. "Ironies of Automation." *Automatica*. Vol. 19, No. 6 (1983) pp. 775–779. http://www.bainbrdg.demon.co.uk/Papers/Ironies.html

Burns, C.M. and Hajdukiewicz, J.R. *Ecological Interface Design.* London, New York et al.: CRC Press, 2004.

Crowl, D. *Human Factors Methods for Improving Performance in the Process Industries.* Hoboken: John Wiley & Sons, Inc., 2007.

EEMUA (Engineering Equipment and Materials Users' Association) *Publication 201: Process Plant Control Desks Utilizing Human-Computer Interfaces—A Guide to Design, Operational and Human Interface Issues.* London: EEMUA, 2002.

Endsley, M., Bolté, B., and Jones, D. *Designing for Situation Awareness.* London: Taylor & Francis, 2003.

Energy Institute, The *Improving Alertness Through Effective Fatigue Management.* London: The Energy Institute, 2006.

IEC 61511-1 *Functional Safety—Safety Instrumented Systems for the Process Industry Sector.* Geneva: International Electrotechnical Commission, 2003.

ISO International Standard 9241—*Ergonomics of Human-System Interaction.* Parts 1–410, Geneva: International Standards Organization, 2008.

ISO *International Standard 11064—Ergonomic Design of Control Centers, First Edition, Parts 1–8.* Geneva: International Standards Organization, 1999.

Johannsen, G. *Mensch-Maschine-Systeme.* Berlin: Springer, 1993.

Kirwan, B. and Ainsworth, L.K. *A Guide to Task Analysis.* London, New York et al.: CRC Press, 1992.

Microsoft Corporation *Microsoft Windows User Experience: Official Guidelines for User Interface Developers and Designers.* Redmond: Microsoft Press, 1999.

Miller, G.A. "The Magical Number Seven, Plus or Minus Two: Some Limits on our Capacity for Processing Information." *The Psychological Review*, Vol. 63, Issue 2, pp. 81–97, 1956.

Pretlove, J. and Skourup, C. "Human in the Loop." *ABB Review*, 1/2007, pp. 6–10. Available under http://bit.ly/17uqb7 (last retrieved 21.05.2009).

Rasmussen, J. and Vicente, K. "Coping with Human Errors Through System Design: Implications for Ecological Interface Design." *International Journal of Man-Machine Studies* 31(5) (1989) pp. 517–534.

Rasmussen, J. "Skills, Rules, and Knowledge; Signals, Signs, and Symbols, and Other Distinctions in Human Performance Models." *IEEE Transactions on Systems, Man, and Cybernetics*, 13(3), May/June 1983, pp. 257–266.

Reason, J. *Human Error.* New York: Cambridge University Press, 1992.

Wickens, C.D. and Hollands, J.G. *Engineering Psychology and Human Performance*. Upper Saddle River: Prentice-Hall, 2002.

6.2 Alarm Management

Martin Hollender

Alarm Systems

The concept of alarms (French for "a l'arme" which means "to the weapons") is very old and originates from the military concept where a guard warns his fellows in the event of an attack. Also, in process control, alarms have a very long tradition (e.g., a whistle indicating that water is boiling). Older plants used panels with light bulbs and bells to alarm the operators. The intuitive understanding is that in case of an alarm the human operator should take some action. From that perspective, alarms are an important boundary between automation and a human operator.

The supervision of a process is usually supported by an alarm system. Theoretically every single alarm should be valuable to the operator and require operator action (see EEMUA 191, 2007). A false negative alarm is defined as the absence of an alarm when a valid triggering event has occurred, and a false positive alarm as the presence of an alarm when no valid triggering event has occurred (EEMUA 191 calls this a nuisance alarm). These two types of failures determine the quality of an alarm system. The alarm system quality results both from the quality of the physical components (sensors) and even more from the effort spent to engineer the alarm system. A perfect alarm system would have neither false positive nor false negative alarms. In practice, a low false negative alarm rate is often paired with a high false positive alarm rate.

Recent alarm system research has focused on eliminating false positive alarms. EEMUA 191 has emphasized the fact that the human capacity to absorb alarms is limited. If this limit is exceeded for longer periods of time, it is likely that important alarms will be overlooked, and in extreme cases the whole alarm system might be more or less ignored.

Figure 6–5 shows the supervisory control task of the operator with a focus on the support given by the alarm system. It is obvious that operators

6.2 *Alarm Management* 379

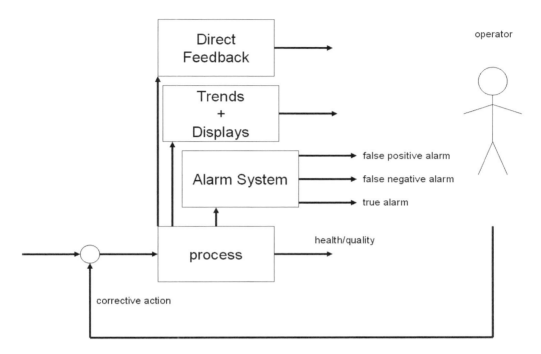

Figure 6–5   Overall Human-Machine System

have many more tasks than just reacting to alarms. Those tasks are interrupted and disturbed by alarms. In the case of true alarms this is inevitable, but in the case of nuisance alarms these interruptions add to the cost of poor alarm system quality. Too-frequent interruptions can also affect the operators' health.

It is a common finding that if there are too many audible alarms, they are turned off. In fact, many control rooms have turned off audible alarms completely, because otherwise the noise level would be intolerable. Alarm rates that are constantly too high distract operators, decrease vigilance and trust, lower situational awareness, and overload the operators' short-term memory. As one consequence, operators may overlook useful alarms that are an early indication of upcoming problems. Large numbers of nuisance alarms decrease the value of important alarms, because the operators are less able to absorb important alarms. Automated systems may be underutilized or disabled because of distrust, as in the case of alarm systems that frequently give false alarms (Parasuraman, Sheridan and Wickens, 2000).

In many control rooms one can easily find symptoms of low alarm system quality like:

- Screens are always covered with alarms, new alarms coming in with high speed (both in normal operation and even more after plant upsets).
- Alarms standing for long periods (days, weeks, and even years).
- Alarms are acknowledged in bulk and without further thought ("blind acknowledgement").
- Operators don't value the alarm system as support for their tasks.
- Audible alarms are disabled because otherwise they would pollute the control room with constant noise.

Why Plants Generate so Many Alarms

With modern control systems, including CPAS, it has become very easy to configure large numbers of isolated alarms. If an approach is taken that does not take the whole system into account, but configures the alarms device by device and signal by signal, this can lead to a cacophony of alarms, diminishing the value of true alarms requiring operator action. Systems may generate lots of alarms (more than 2000 alarms per day is typical for many plants) during normal operation, and even more during process upsets.

Modern fieldbus devices can generate a broad range of diagnostic messages, which are often issued as alarms. Modern control systems allow configuration of the target user group for alarms and events, so that maintenance messages without direct relevance for operators are only presented to maintenance staff. However, such messages are often simply sent to everyone, including the operators.

Malfunctioning sensors like dirty position indicator switches sometimes generate high rates of ON-OFF-ON alarms. It is not always possible to quickly replace or repair those sensors because they might be inaccessible. If such alarms are disabled, proper Management of Change needs to be in place, because otherwise it might be forgotten to enable the alarm again once the sensor has been repaired.

The configuration of many control loops often is far from optimal (see Section 5.3). Such loops can drive actuators in extreme positions which might

6.2 Alarm Management

result in alarms. Many function blocks for control loops have diagnostic functions included which can generate alarms. A systematic analysis of alarms related to control loops can be used to identify candidates for further loop monitoring (Section 5.3) or loop tuning (Section 5.2).

Device manufacturers try to guard themselves against warranty claims by setting lots of tight alarms. This might make sense from the perspective of the local device, but it contributes to alarm overload and therefore lowers overall alarm system quality.

Often special alarms are configured during commisioning phase. Once commissioning is done, many of those alarms make no sense any longer but are never removed.

Today's control systems allow the configuration of several different alarms for each and every tag; for example:

- Low and high limit alarms with 3 levels each (L,LL,LLL,H,HH,HHH)
- Rate of change, decreasing and increasing
- Target Deviation Alarms
- Bad Value Alarms

Alarm Management Guidelines and Standards

A thorough analysis of accidents like the explosion in the Texaco Refinery, Milford Haven (Health and Safety Executive, 1997) has clearly shown that bad alarm management has contributed to accidents. At Milford Haven, the operators got 275 different alarms in the last 11 minutes before the explosion. This is why authorities like HSE (Health and Safety Executive) in the UK or the Norwegian Petroleum Directorate require safety-critical plants to implement systematic alarm management.

If critical situations can be avoided by operators stabilizing the process so that the emergency shutdown system doesn't need to be used, this not only increases the safety of the plant, but has in many cases substantial economic benefits, because every unplanned shutdown can be very expensive. The increased safety is the visible tip of the iceberg, but better alarm management helps to run the process better.

The most influential document for alarm management so far has been the EEMUA 191 guideline. ISA has recently published ANSI/ISA-18.2-2009.

Other important guidelines for alarm management are NAMUR NA102 (2008) and the Norwegian YA 711 (Norwegian Petroleum Directorate 2001).

EEMUA 191

In 1999, the Engineering Equipment and Materials Users Association (EEMUA) produced its Publication 191, "Alarm Systems, a Guide to Design, Management and Procurement." EEMUA is a non-profit industry association run for the benefit of companies that own or operate industrial facilities. EEMUA 191 has become the world-wide de facto standard for Alarm Management. Key ideas in the guideline are that every alarm should be useful and relevant to the operator and that there should be no alarm without a predefined operator response.

The common expectation that operators should never overlook important alarms is impractical if new alarms arrive too rapidly, day by day, 8 hours a shift. EEMUA 191 recommends that the alarm rate in normal operation should be below 1 alarm in 10 minutes.

EEMUA 191, which was updated in 2007, is a guideline for Alarm Management and is not mandatory. The document is recognized by a number of regulatory bodies as good practice. EEMUA 191 focuses on the operator's information processing capabilities. It emphasizes the usability of an alarm system from the operator's perspective. The document is only available on paper and can be bought on www.eemua.org.

ANSI/ISA-18.2

For several years, an ISA committee has been focusing on the development of the standard titled ANSI/ISA-18.2, "Management of Alarm Systems for the Process Industries." The committee has wide representation from users, vendors, and consultants. As previously mentioned, the standard was published this year. As ANSI/ISA-18.2 incorporates work and ideas from EEMUA 191 and is a standard (not a guideline, as EEMUA 191), it is likely that ANSI/ISA-18.2 will take over the role of EEMUA 191 as the world-wide standard for alarm management.

The scope of ANSI/ISA-18.2 is to establish terminology and practices for alarm systems, including definition, design, installation, operation, maintenance and modification, and work processes recommended to effectively maintain an alarm system over time. Alarm system management includes multiple work processes throughout the alarm system life cycle.

NAMUR NA 102

NAMUR (International User Association for Automation in Process Industries) worksheet NA 102 sets out a procedure for designing alarm management within a process control system, starting from a holistic view of the process. While it does include message signals, its main focus is on alarms. During the engineering and erection phase, the worksheet is intended to be used as a general guide, based on which a part of an individual PAS specification can be derived by tailoring it to fit a particular unit of equipment. During plant operation, it serves as a guide for service and maintenance as well as signaling system optimization. It also gives information for manufacturers of process control systems on how to extend product functions. NA 102 describes one example plant where improved alarm management resulted in increased overall equipment effectiveness (OEE) by more than five percent.

Plant-wide Alarm Philosophy

A plant-wide alarm philosophy document describes how alarms should work in a plant. It is very important to have one written document describing a consistent plant-wide alarm philosophy. Details can be found in EEMUA 191. Items in the document should include:

- Required alarm system quality. Target values for Key Performance Indicators (KPIs) and when and how these KPIs are measured.
- Name of the manager responsible for the quality of the alarm system.
- Selection criteria for what is alarmed and what is not.
- The different priorities used and their associated meaning; rules for how alarm priorities are systematically assigned.
- Management of Change for alarm configuration parameters.

Alarm System Harmonization

In plants, different parts are commonly automated at different times, with different systems by different companies. The philosophies behind the alarm systems for those different parts can be rather different. As it might happen that a single operator is responsible for supervising different parts of the plant, inconsistencies can be very confusing and present a considerable risk. For example, if the priority value 1 means high in one part of the plant and low in another, this can be dangerously misleading. The CPAS must support the harmonization of different alarm sources by allowing the re-mapping of priority values (see Section 4.1 on alarms and events) because for technical or organizational reasons it is not always possible to harmonize the alarm settings directly at the source.

Alarm System Quality Metrics

In the past it was difficult to assess the quality of an alarm system. When modern technology began to generate more and more alarms, many plants accepted excessively high alarm rates. Investments in alarm systems were difficult to justify because the potential improvements were hard to quantify. It was a big step forward when in 1999 EEMUA specified several measurable performance indicators that can be used to benchmark a plant's alarm system performance, as, for example:

- Long-term average alarm rate in steady operation
- Number of alarms during first 10 minutes after a major plant upset
- Alarm priority distribution: high (5%) – medium (15%) – low (80%)
- Percentage of intervals exceeding a threshold value
- Average number of standing alarms

These easy-to-measure KPIs don't automatically guarantee a good alarm system, but they are useful tools on the way towards better quality (Hollender and Atkinson, 2007).

Alarm System Quality Improvement

The basic steps of alarm system quality improvement are:

- Record all alarms and events into a database as a basis for further analysis. Many control systems include an alarm logging module, or third-party alarm loggers can be added to the system. Alarm printers as still seen in some plants are difficult and expensive to maintain and essentially bury valuable information about the plant's history in huge piles of paper that nobody ever reads.

- Measure alarm rates and other alarm Key Performance Indicators and compare them with recommendations from guidelines or values from reference plants. The overall approach can be decided based on these KPIs (Tanner et. al. 2005).

- Identify "low-hanging fruit"—it is often possible to improve the alarm system significantly with relatively little effort. A few alarms might be responsible for a large part of the overall alarm load.

- Eliminate nuisance alarms. Doing this typically includes:
    - Tuning control loops
    - Adjusting alarm limits, filter settings and hysteresis
    - Replacing faulty sensors
    - Configuring alarms that require no operator action as events, or removing them.

- Alarm Rationalization is the process of reviewing alarms against the principles of the alarm philosophy and determining and documenting the rationale and design requirements for the alarm. This includes the basis for the alarm setting, the consequences of deviation, and corrective action that should be taken by the operator. Rationalization also includes the prioritization of an alarm based on the mechanism defined in the alarm philosophy, or its removal.

- Ensure that alarms are handled timely. Alarms standing for days, weeks, or even years are often an indication of a culture accepting risk and suboptimal quality.

- Measure the alarm key performance indicators regularly to ensure they stay in the desired target area. As the plant changes over time, it is important to establish Alarm Management as part of the plant's procedures. (Alarm Management is often seen in the context of Six Sigma programs.)

Systematic alarm management usually reveals weak spots in the plant and can therefore directly help to improve plant performance. The operator guidance given by a high-quality alarm system will further improve plant performance.

Alarm Configuration Management of Change (MoC)

The following information should be recorded when changes to alarms are approved: reason, date, who made the change, nature of the change, and training taken by those affected. U.S. Department of Labor Occupational Safety & Health Administration regulation 1910.119 (OSHA, 1996) requires safety-critical plants to implement Management of Change procedures, including the alarm system. IEC 61511 (2003) also requires Management of Change (Mannan and West, 2004). Management of Change mechanisms in a CPAS help to systematically risk-assess proposed configuration changes and ensure that only approved changes are implemented. All configuration changes that potentially have an impact on safety should be handled with such mechanisms. Depending how safety-critical a process is, even small, local and temporary changes should be included.

It should be possible to periodically compare the approved alarm configuration parameters with the parameters currently active in the CPAS. This helps to maintain the integrity of the alarm system. All differences between approved and currently active settings are automatically detected and brought to the proper person's attention for review.

Advanced Alarming

In the 1990s many people (including the author) dreamt that advanced alarming mechanisms based, for example, on expert system technology could lead to higher quality alarm systems. In the meantime it has become clear that no silver bullet exists and that there is no way around systematic hard engineering work as described in this chapter. Below, a few more advanced alarming techniques

6.2 Alarm Management

are described. Generally, the simpler an alarm system can be to achieve the desired operational result, the better chance it has of sustained benefit and success. When considering the use of enhanced and advanced alarm systems one should weigh the expected benefits in safety, environmental protection, cost, or other criteria against the increased complexity of system design, implementation, and maintenance (ISA-18.02). It has happened that adding a complex and non-transparent support system has actually increased the operators' alarm burden.

Alarm Addressing

In older systems all alarms were presented to the operators, often in a single alarm list. As has been mentioned, CPASs contain distribution mechanisms which allow the addressing of specific alarms to specific user groups. Each user group, such as system engineers, operators, maintenance staff, and application engineers can have their own dedicated alarm lists showing only the alarms relevant to their role. Specific alarms can be distributed to them by email or Short Message Service (SMS) message. For example, the message that the operating hours of a motor have exceeded a threshold and maintenance should be scheduled is not relevant for operators and therefore it should be sent to maintenance staff only. Usually the assignment of alarms to user groups is done with the help of the event category, but other event attributes can also be used.

Alarm Filtering

CPAS allow to configure many different alarm lists and to define complex filter for each list (see Section 4.1). It must be clearly visible which filter configuration is currently used and only a limited set of consistent system-wide filter settings should be used. If operators would be able to make individual changes to filter settings of alarm lists this will result in dangerous confusion once the alarm list is used by an operator unaware of the filter configuration change.

Alarm Hiding

Alarm hiding is used on process areas or objects to hide alarms which are not of interest at a specific stage or area. Alarm rules can be defined so that certain alarms will be hidden, dependent on process state or other active alarms.

Hidden alarms have a reduced visibility but are still available if explicitly requested (see also Section 4.1 on alarms and events).

Alarm Shelving

Shelving is a facility where the operator is able to temporarily prevent an alarm from being displayed when it is causing a nuisance. A shelved alarm will be removed from the list and will not re-annunciate until un-shelved (from EEMUA 191). Even in alarm systems with a very high quality when operating under normal circumstances, high numbers of recurring alarms may be created due, for example, to faulty instruments or devices under maintenance. Shelving is a mechanism to temporarily work around this problem. Disabling such an alarm altogether would incur the risk that the alarm is not re-enabled once the problem is fixed.

It is important to ensure that shelved alarms are never forgotten and that all colleagues and members of other shifts affected by the alarm are aware of the fact that the alarm has been shelved. This can be done by forced reviews of all shelved alarms and by associating a timer with each shelved alarm. All shelved alarms should be periodically reviewed.

Alarm Grouping

A single device can generate several causally interrelated alarms. Sometimes only one of those alarms will be activated (e.g., temperature too high), but in the case of a bigger problem, several alarms will be activated. For example, if a compressor has tripped, detailed alarms from the internals of the compressor are only of interest when specifically diagnosing this one compressor. In the general alarm list, one alarm indicating that the compressor has tripped is sufficient and any additional, more detailed alarms would only distract the operators' attention. Any more detailed compressor alarms (in this example) should be grouped under the general trip alarm and only show up if specifically requested.

Alarm Load Shedding

As we have seen, the human capacity to absorb alarms is limited. If the alarm rate is higher than this limit, the current strategy of many operators is to ignore

6.2 Alarm Management 389

some or even all alarms. EEMUA 191 discusses a research concept termed automatic alarm load shedding. The idea is that the alarm system automatically limits the alarm display rate to a rate an operator can cope with. One criticism of this approach is that it is probably not always transparent to the operator why the alarm system presents different alarms at a certain point. A better alternative implemented in many plants may be to use manual alarm load shedding. This means that the operator can select a predefined filter that shows the most important alarms only. Such a filter depends on a good classification of the alarms; for example, if alarm priorities are used, the most important and urgent alarms would have the highest priority assigned.

Suppression of Alarms from Out-of-service Areas

In many plants some of the nuisance alarms come from process areas that are either currently not used or are under maintenance. A CPAS allows the hiding or disabling of such alarms (see Section 4.1 on alarms and events). It must be ensured that the alarms will be visible again once the maintenance is finished. Therefore these changes to the alarm configuration must be part of the Management of Change system. If out-of-service areas can be determined by the Computerized Maintenance Management System (CMMS), this information can be used to hide the corresponding alarms.

State-based Alarms

The standard practice in current CPAS managing continuous processes is to use static alarm thresholds. In some cases it would make sense to use flexible thresholds, depending on the current situation. For example, the threshold for the bearing temperature can be made dependent on the outdoor temperature because on a cold winter day, the thresholds can be set lower. In winter an alarm temperature of 90°C might work well, but in summer the same alarm temperature might result in lots of false alarms. Of course CPAS allow calculating alarm thresholds based on physical formulas which would make sense from a scientific point of view, but in practice this is rarely done.

In addition, many alarms are designed for steady state only, but often start-up and shut-down are the most difficult phases from an operational point of view, and during these times the operators require the least distraction. Alarms that only make sense in steady state should be hidden during start-up and shut-down. States can be determined through a calculated variable or by operator indication (see Section 4.2). This is also valid for batch plants, especially those that run many recipes (see EEMUA 191).

References

EEMUA 191 *Alarm Systems. A Guide to Design, Management and Procurement, 2nd Edition.* London: EEMUA, 2007. http://www.eemua.org.

Health and Safety Executive, *A report of the investigation by the Health and Safety Executive into the explosion and fires on the Pembroke Cracking Company Plant at the Texaco Refinery, Milford Haven on 24 July 1994.* Norwich: Health and Safety Executive, 1997. ISBN 0717614131.

Hollender, M. and Atkinson, T. "Alarms for Operators." *Proceedings OECD-CCA Workshop on Human Factors in Chemical Accidents and Incidents.* Potsdam, 2007.

IEC 61511 *Functional Safety—Safety Instrumented Systems for the Process Industry Sector.* Geneva: IEC, 2003.

ANSI/ISA-18.2 *Management of Alarm Systems for the Process Industries.* Research Triangle Park: ISA, 2009. http://www.isa.org.

Mannan, H. and West, H. "Management of Change of Chemical Process Control Systems." Paper presented at the International Conference on the 20th Anniversary of the Bhopal Gas Tragedy. Kanpur, India, 2004. Available at http://bit.ly/zMZQQ (last retrieved 16.5.2009).

NAMUR NA102 *Alarm Management.* NAMUR, 2008. http://www.namur.de

Norwegian Petroleum Directorate YA 711 *Principles for Alarm System Design.* Stavanger: NPD, 2001. Available at www.npd.no and bit.ly/1tFPOh (last retrieved 16.5.2009).

OSHA *Regulation 29 CFR 1910.119 Process Safety Management of Highly Hazardous Chemicals.* Washington, DC: U.S. Occupational Safety and Health Administration, 1996. Available on the Internet at www.osha.gov.

Parasuraman, R., Sheridan, T.B., and Wickens, C. "A Model for Types and Levels of Human Interaction." *IEEE Transactions on Systems, Man, and Cybernetics,* Vol. 30, No. 3 (2000).

Tanner, R., Gould, J., Turner, R., and Atkinson, T. "Keeping the Peace (and Quiet)." *InTech*, Research Triangle Park: ISA, Sep. 2005. Available at bit.ly/Q2c0E (last retrieved 16.5.2009).

List of Acronyms

APC	Advanced Process Control
API	Application Programming Interface
ASCII	American Standard Code for Information Interchange
B2MML	Business To Manufacturing Markup Language
BPCS	Basic Process Control System
CAE	Computer-Aided Engineering
CIM	Computer Integrated Manufacturing
CMMS	Computerized Maintenance Management System
COM	Component Object Model
CORBA	Common Object Request Broker Architecture
COTS	Commercial Off The Shelf
CPAS	Collaborative Process Automation System
CPM	Collaborative Production Management
CPU	Central Processing Unit
CSV	Comma Separated Values
DCOM	Distributed COM
DCS	Distributed Control System
DoS	Denial-of-Service
DTM	Device Type Manager
EAM	Enterprise Asset Management
EDDL	Electronic Device Description Language

EEMUA	Engineering Equipment & Materials Users' Association (www.eemua.org)
EID	Ecological Interface Design
EPC	Engineering, Procurement & Construction
ERP	Enterprise Resource Planning
FAT	Factory Acceptance Test
FBD	Function Block Diagram
FDA	U.S. Food and Drug Administration (fda.gov)
FDT	Field Device Tool
FEED	Front-end Engineering and Design
FF	Fieldbus Foundation
GUI	Graphical User Interface
HART	Highway Addressable Remote Transducer Protocol
HCI	Human Computer Interaction
HMI	Human-Machine Interface
HSE	Health and Safety Executive
HSI	Human System Interface
HTML	Hypertext Markup Language
HTTP	Hypertext Transfer Protocol
I&C	Instrumentation & Control
IDM	Intelligent Device Management
IEC	International Electrotechnical Commission
IETF	Internet Engineering Task Force
IO	Input/Output
ISA	International Society of Automation
ISO	International Organization for Standardization

List of Acronyms

IT	Information Technology
KKS	Kraftwerk-Kennzeichen-System
KPI	Key Performance Indicator
LAN	Local Area Network
LIMS	Laboratory Information Management System
MES	Manufacturing Execution System
MILP	Mixed Integer Linear Programming
MIMOSA	Alliance of Operations & Maintenance (O&M) solution providers and end-user companies (mimosa.org)
MMI	Man-Machine Interface
MPC	Model Predictive Control
MTBF	Mean Time Between Failures
NAMUR	International user association of automation technology in process industries (namur.de)
NLS	Native Language Support
NMPC	Nonlinear Model Predictive Control
ODBC	Open Database Connectivity
OEE	Overall Equipment Effectiveness
OLE	Object Linking and Embedding
OMAC	Organization for Machine Automation and Control
OPC	Previously OLE for Process Control; now simply OPC
OpX	Operational Excellence
P&ID	Piping and Instrumentation Diagram
PAM	Plant Asset Management
PAS	Process Automation System
PID	Proportional/Integral/Derivative

PIMS	Plant Information Management System
PLC	Programmable Logic Controller
RAID	Redundant Array of Independent Disks
RBAC	Role-based Access Control
RDBMS	Relational Database Management System
REST	Representational State Transfer
RGB	Red-Green-Blue Color Model
RMI	Remote Method Invocation
SBC	State Based Control
SCADA	Supervisory Control And Data Acquisition
SCM	Supply-Chain Management
SDCV	Spatially Dedicated Continuously Visible
SDK	Software Development Kit
SIF	Safety Instrumented Function
SIL	Safety Integrity Level
SIS	Safety Instrumented System
SIT	System Integration Test
SMS	Short Message Service
SNMP	Simple Network Management Protocol
SOA	Service-Oriented Architecture
SOAP	Previously Simple Object Access Protocol; now simply SOAP
SPC	Statistical Process Control
SRK	Skill-, Rule-, and Knowledge-based
SQL	Structured Query Language
TCO	Total Cost of Ownership
TMR	Triplicated Modular Redundancy

List of Acronyms

TQM	Total Quality Management
UML	Unified Modeling Language
VPN	Virtual Private Network
W3C	World Wide Web Consortium
WBF	Previously World Batch Forum; now The Organization for Production Technology
WSDL	Web Services Description Language
XML	Extensible Markup Language
XSD	XML Schema Definition
XSLT	Extensible Stylesheet Language Transformation

Index

A

access control
  statements   126
acknowledgment   216
actionable information   36
ActiveX   76
activity based costing   38
actuators   3
Ad-Hoc decision making   36
administration   50
Advanced Process Control (APC)   244
alarm indicators   213–215
  alarm and event lists   213
  alarm bands   213
  alarm historians   215
  alarm printers   214
  alarm sequence bars   213
  audible and external alarms   215
  email   215
  notification to external systems   215
  SMS (Short Message Service)   215
  spatial dedicated alarm displays   214
alarm systems   210
  analysis   215
  categories   222
  chattering   213
  management   240
  priorities   220
  vendor-specific attributes   221
algorithms
  deadband   300
  evolutionary   334
  swinging door   300
ANSI/ISA-99.00.01-2007   119
APC (Advanced Process Control)   244
application
  interfaces   306–307

separation   123
servers   22, 32
architecture
  alternatives   142
  custom developments   67
  flexibility   67
  high-integrity   113
  integrated   142
  interfaced   142
  separate   142
ASCII   70
Aspect Object Model   44
asset
  management   281
  optimization   281
authorization   75
automation
  irony of   355
  levels   376
  pyramids   3
  system engineering   157
availability   100, 137

B

B2MML. *See* Business to Manufacturing Markup Language
back-ups
  standby   104
  workby   104
base types   87
batch processes   3
best operator   34
best practices procedure   230
best-in-class components   68
binary formats, custom   71
binary software component   74
boiler lifetime monitoring   301

bottlenecks   115
Branch-and-Bound   339
brown field   160
bump test   261
Business To Manufacturing Markup Language (B2MML)   69, 84, 319, 320

C

cable planning   165
calculations   301–302
cement   16–17
centers of excellence   238
centralized control room   241
change control   151–153
chattering alarms   213
    See also alarm systems
chemical   13–14
    petrochemicals   13
CIM (computer-integrated manufacturing)   26, 315
CMMS. *See* computerized maintenance management systems
cogeneration of steam and power   249–250
cognitive biases   358
collaboration   353
Collaborative Process Automation System. *See* CPAS
COM (Component Object Model)   74
comma separated values (CSV)   71
commercial-off-the-shelf (COTS)   118
commissioning   172
    cold   172
    hot   172
Common Information Infrastructure   27, 32
communications
    infrastructure   88
    protected   123
compliance
    certificate   89
    program   89
    test tools   89
Component Object Model (COM)   74
compression methods   300

computer-integrated manufacturing (CIM)   26, 315
computerized maintenance management systems (CMMS)   4, 215, 288, 325
condition monitoring   286
configurable solutions   116
control
    logic engineering   169
    loop oscillations   270
    loop variance   270
    model-based   244
    modules   227
    performance   270
control room design
    furniture   364
    ISO 11064   363
    task zones   363
controllers
    automation   47
    connectivity   47
    fail-over logic   104
    fail-safe   104
    fail-silent   103
copper plant scheduling   342–343
COTS (commericial-off-the-shelf)   118
CPAS   25
    advanced applications   36–38
    architecture   42
    containment   123
    costing   116
    customized layouts   116
    emergence   29
    guiding principles   29
    remote troubleshooting   238
    security   122
    service frameworks   46
    traceability support   149
CSV (comma seperated values)   71
cycle times   21

D

dangerous or hostile locations   241
data
    acquisition interfaces   298

long-term storage   298
quality   32
reconciliation   311
restricted flows   122
data collectors
    redundant   297
    store-and-forward   297
DCOM   46, 75, 95
DCSs (Distributed Control Systems)   21
decoupled operators   376
Denial-of-Service (DoS)   120
dependability   100
deterrence   128
Device Type Manager (DTM)   79
DeviceNet   32
digital fieldbuses   19, 77
digital or analog input cards   19
digital signatures   129
discovery mechanisms   89
distributed garbage collection   75
disturbance analysis   276
DMZ. *See* demilitarized zone
documentation   172
DoS (Denial-of-Service)   120
DTM (Device Type Manager)   79

E

Ecological Interface Design (EID)   369
EDDL (Electronic Device Description Language)   79, 94
EEMUA
    191   220
    201   365, 369, 374
EEMUA (Engineering Equipment & Materials Users' Association)   210
EID (Ecological Interface Design)   369
electric power   10
electrical components
    integration of   82
Electronic Device Description Language (EDDL)   79, 94
electronic records   149
electronic signatures   149
EM (Equipment Modules)   227

energy management   310, 327
engineering   156
    efficiency   231–236
Engineering Equipment & Materials Users' Association (EEMUA)   210, 220
enterprise connectivity   313–328
    framework   321
Enterprise Resource Planning (ERP)   298, 313
environmental regulations   17
EPC (engineering, procurement and construction)   157
Equipment Modules (EM)   227
ERP (Enterprise Resource Planning)   298, 313
error detection   104
error masking   107
ethernet   26
event
    condition-related   211
    history   93
    simple   211
    time-stamped audit trail   152
    tracking-related   211
expert system technology   34
extensibility   91
Extensible Markup Language (XML)   69–72
    schema   69
Extensible Stylesheet Language Transformation (XSLT)   69

F

Factory Acceptance Test (FAT)   171, 172
failover   103
fault diagnosis   240
fault-tolerant   102
    automation systems   100
    operation   32
FDA 21 CFR Part 11   147
FDI (Future Device Integration)   79
FEED (front end engineering and design)   158
Field Device Tool (FDT)   78, 94

field device, configuration   78
field hardware test   172
field instrumentation   159
file system integrity checkers   129
filtering   218
firewall-friendliness   305
food and beverage   12
    food safety   12
Forum for Automation and Manufacturing Professionals (WBF)   319
FOUNDATION Fieldbus   32, 77
front end engineering and design (FEED)   158
functional view   30–32
Future Device Integration (FDI)   79

G

Gantt chart   336
globalization   238
    advantages of   238
    reduced downtime   238
    remote service   238
green field   159

H

handover   172
hardware
    field testing   172
    random failure   135
Harris Index   273
HART (Highway Addressable Remote Transducer)   77
HCI (Human Computer Interaction/Interface)   353
heuristics   334, 358
hiding   218
hierarchical logs   300
Hierarchical Task Analysis (HTA)   363
hierarchies   91
high-integrity system   103–104
Highway Addressable Remote Transducer (HART)   77
HMI. *See* Human-Machine Interface

HSI (Human System Interface)   354, 356
    guidelines   370
    Window system guidelines   371
HSI design   364–370
    alarm system   367
    big picture   368
    checklists   367
    deviation from normal operation   367
    device status displays   367
    graphical libraries   366
    keyhole effect   366
    large screens   368
    look and feel   365
    Microsoft Windows model   364
    number of screens   365
    operator participation   369
    operator requirements   366
    Piping and Instrumentation Diagram (P&ID)   368
    problem diagnosis   367
    process interlocks   367
    Safety Integrity System (SIS)   368
    task-based graphics   367
    team-based communication   368
    trend displays   367
    trend plots   368
    unit level mass balances   368
    unit status displays   367
    usability engineering   370
HSL design for color-blind operators   366
HTA (Hierarchical Task Analysis)   363
HTML   72
HTTP   72
Human Computer Interaction/Interface (HCI)   353
    undo-button   354
human error   358
    cognitive biases   358
human operator
    characteristics of   358–359
    control room operator   355
    creativity   376
    flexibility   376
    goals   359, 360
    human short-term memory

capacity   358
safety layers   360
situation awareness   359
vigilance   360
Human System Interface (HSI)   354, 356
Human-Machine Interface (HMI)   354
configuration   168
hybrid control   254

I
I/O
communication setup   167
handling   165
I&C   159
IDM (Intelligent Device Management)   240
IEC
11064   369
61131   94
61131-3   82
61346   43
61508   102, 133
61511   102, 133
61512   84
61850   82, 241
62061   133
62264   83, 315
IEC/EN 61508   83
IEEE
802.3   26
802.4   26
independent test labs   89
information
distribution   303–306
management   294
modeling   90–93
information models. *See* models
installation   172
instrumentation & control   159
integrated re-trimming module   344–346
integration   138
integrity breach   101
Intelligent Device Management (IDM)   240
internet   72–74, 118
interoperability workshops   89

interoperable models. *See* models
irony of automation   355
ISA-88   28, 84, 224
ISA-95   4, 315, 319, 348
ISO 11064   365, 373
ISO 13849   133
ISO 9241   370, 371
ergonomics   371
IT security   118
risk management   119–120

J
Jacquard Loom   18
job design
analysis phase   363
competence profiles   356
diagnosis   363
emergency response coordination   363
manning levels   356
optimization   363
permits to work   363
shutdown   362
startup   362

K
Kalman filter   246
Key Performance Indicators (KPI)   308
knowledge workers   4
kurtosis   272

L
Laboratory Information Management
   Systems (LIMS)   298
ladder logic   19
Lambda Tuning   262–266
late binding technology   31
least privilege   125
LIMS (Laboratory Information Management
   Systems)   298
linear programming   338
Linux   46
logical independence   138
logical view   32–34

loop
- monitoring   269
- tuning   240, 258, 267

M

Machinery Information Management Open Systems Alliance (MIMOSA)   94
maintainability   100
maintenance   283
- predictive   285, 289
- preventive   285, 288
- proactive   285
- reactive   285, 288
- remote   238
- strategy   284

malware   124
Man-Machine Interaction/Interface (MMI)   354
Manufacturing Execution Systems (MES)   313
manufacturing operations
- benchmarking   35
- critical situation management   36
- operator training   35
- performance management   35
- work process management facility   35

manufacturing service bus   307
marshalling   75
master terminal unit (MTU)   22
master time   34
material and energy balance   311
MES (Manufacturing Execution Systems)   313
Microsoft Windows   46
MILP (mixed integer linear programming)   339
MIMOSA (Machinery Information Management Open Systems Alliance)   94
mining and metals   15–16
- open pit mining   15

mixed integer linear programming (MILP)   339
mixed logic and dynamic (MLD)   254
MMI (Man-Machine Interaction/Interface)   354
model predictive control (MPC)   244
models
- black-box   245
- hybrid   245
- information   87
- interoperable   87
- Purdue   324
- standard information   94
- white-box   245

modes of operation   360
- checklists   362
- diagnosis   362
- normal   360
- planned abnormal   360
- planned maintenance   361
- procedures   362
- unit shut-down   361
- unit start-up   361
- unplanned abnormal   360

Modicon   19
modular design   115
modularity   138
monitoring & audit   123
moving horizon estimation   247
MPC (model predictive control)   244
MTU (master terminal unit)   22

N

network
- intrusion detection systems   129
- separation   122

Nichols   259
NLP (nonlinear programming)   249
NMPC (nonlinear MPC)   249
node class
- method   91
- object   91
- variable   91

nodes   91
- attributes   91
- view   93

nonlinear MPC (NMPC)   245

nonlinear programming (NLP)   249
notifications   305

O

object
    lookup   48
    management   28, 30
object instances   45
object types   45
    proven best-in-class solutions   45
    standardized solutions   46
objective function   249, 334
object orientation   43–45
object-oriented techniques   87, 90
OEE (overall equipment effectiveness)   308
offer preparation   163
oil and gas   11–12
    downstream   11
    liquefied natural gas (LNG)   12
    marine   11
    pipelines   11
    upstream   11
OLE for Process Control (OPC)   80
    clients   80
    model
        UA meta   88
    OPC AE   81, 210, 306
    OPC Classic   80
    OPC DA   81, 306
    OPC HDA   82, 306
    server   80
    Unified Architecture (OPC UA)   80, 86
    XML-DA   86
one size-fits-all   113
OPC. *See* OLE for Process Control
Open Modular Architecture Controls (OMAC)   94
operational complexity
    cognitive requirements   353
operations
    abnormal   362
    cognitive requirements   353
    complexity   353
    consistency   229–231
    deviations   362
    excellence   269
    normal   360
    routine   360
operations environment   356
    alarm systems   356
    alertness   357
    competence   356
    control room design   356, 363
    fatigue   357
    job design   356
    procedures   357
    safety critical communications   357
    training   356
operator requirements
    color palette   366
    consistency   366
    keyboard   364
    mouse   364
    patrolling   367
    screens   364
    standard operating procedure   367
operators   364
    cost of downtime   352
    effectiveness of expert users   355
    effectiveness of human operators   352
    empowerment of   4
    jobs   355
    owner/operator   157
    training   35, 355, 376
optimization   332
    mathematical   334
    production   245
    pulp mill   253–254
    start-up   302
oscillation
    detection   274
    diagnosis   271
overall equipment effectiveness (OEE)   308

P

PAM. *See* Plant Asset Management
PAS. *See* process automation systems

performance
    measurements   272
    real-time management   37
persistency breach   101
PFD (probability of failure to perform a safety function)   134
PFH (probability of failure per hour)   134
pharmaceuticals, biotechnology   12–13
PID controller   258
PIMS (Plant Information Management System)   294
planning and scheduling   330–348
Plant Asset Management (PAM)   28, 281–292
Plant Information Management System (PIMS)   294
plant variability   271
platform-independent   86
PLC (Programmable Logic Controller)   19–21
PLCopen   21, 94
plug-and-play connectivity   49
point-to-point connections   239, 317
power boiler start-up   250–253
power trading   11
probability of failure per hour (PFH)   134
probability of failure to perform a safety function (PFD)   134
procedures
    checklists   357
    consistency of operation   357
process
    incidents   356
    models   9
    states   224–226
process automation systems (PAS)   67
    architectural flexibility   67
    cost of ownership   67
    evolutionary enhancements   67
    interchangeability   68
    interoperability   68
    mixed suppliers   68
    proprietary technology   67
    standards   68

process control
    event monitoring   35
    operational perspective   34
    performance monitoring   34
    procedural best practices   34
    progress and performance visualization   34
    troubleshooting   34
Process Data Historians   26
processes
    automation market   9
    batch   3, 9
    continuous   3
    discrete manufacturing   9
procurement & logistics   166
production
    continuous   9
    optimization   245
PROFIBUS   32, 77
Programmable Logic Controller (PLC)   19–21
project library creation   165
project management   165
protections   101
pulp and paper   14–15
punch cards   18

R

RBAC (role-based access control)   126
redundancy   100, 103
references   91
relational database management systems (RDBMS)   76
reliability   100
remote access
    access management   239
    security   239
remote operation   241
remote terminal units (RTUs)   22
renewable energy   11
reporting   302–303
requirements collection   164
response times   4
REST (Representational State Transfer)   73

Index

return to normal (RTN)   210, 216
risk analysis   119
role-based access control (RBAC)   126
RTN (return to normal)   210, 216
RTUs (remote terminal units)   22

S

safe state   101
safety   100, 133
    critical communications   357
    functional   134
    integrity measures   138
    key components   134
    measurements of   134
safety instrumented systems (SIS)   136
safety integrity level (SIL)   134
SAT (site acceptance test)   172
SBC (State Based Control)   224
SCADA (Supervisory Control and Data Acquisition)   22
scalability   112
    potential for growth   114
    standardization   113
scheduling   335–342
SDCV. *See* situation awareness
security   88, 239
    availability objective   120
    balanced defensive strength   121
    confidentiality objective   120
    defending the network   122
    defense in depth   121
    detecting attacks   121
    false alarms   360
    integrity objective   120
    multiple rings of defense   121
    by obscurity   121
    patch management   240
    principle of least privilege   121
    unauthorized access   126
    vigilance   360
sensors   3
servers
    application   22, 32
    capabilities   89
    CPAS   239
    umbrella OPC   48, 49
service framework   46–47
service-oriented architecture (SOA)   73
services
    application   51
    connectivity   51
    object   51
set point (SP)   246
shelving   217
shutdowns   355
SIL (safety integrity level)   134
Simple Network Management Protocol (SNMP)   50, 73
Simple Object Access Protocol (SOAP)   73, 95, 321
simplex
    method   339
    system   107
single version of the truth   36
SIS (safety instrumented systems)   136
SIT (System Integration Test)   171
site acceptance test (SAT)   172
situation awareness
    large screen displays   359
    SDCV   360
skewness   272
skill-, rule-, & knowledge-based behavior (SRK)   358
smart phones   371
SNMP (Simple Network Management Protocol)   50, 73
SOA (service-oriented architecture)   73
SOAP (Simple Object Access Protocol)   73, 95, 321
SP (set point)   246
SQL (Structured Query Language)   76, 306
staff reductions   376
staging   167
standard application view   34–36
State Based Control (SBC)   224
static friction (stiction)   271
steam engine   18
step response experiment   261
Structured Query Language (SQL)   76, 306

Supervisory Control and Data Acquisition
(SCADA)   22
surface of attack   124
system
    administration   167
    defense   121
    duplex (redundant)   107
    error masking   107
    failure   103
        integrity breach   101
        persistency breach   101
    hardening   124
    high-integrity   103–104
    integration   170
    management   33
    mappings   316
    multisystem integration   241
    persistent   104–107
    remote service   238
    scalability   113
    security   121
    single point of entry   87
    tag-based   31
    versatility   113
System Integration Test (SIT)   171
systematic failure   135, 136

T

task analysis   369
TCP-IP   26
terminal servers   123
thick client applications   295
thin client technology   295
TMR (Triplicated Modular
    Redundancy)   108
traceability   147
triplication   108
troubleshooting   238
tuning
    model-based interactive tools   266–267

relays   261

U

UA Analyzer   87, 94
UA TCP   96
UA devices   87
unicode   70
USB sticks   124

V

VDI/VDE 2182   119
vendor lock-in   68
versatility   112
virus   124
    scanners   129
voting   108
VPN (Virtual Private Network)   122

W

W3C (World Wide Web Consortium)   69
water and wastewater   16
WBF (Forum for Automation and
    Manufacturing Professionals)   319
web services   73
    technology   86
wet ink   149
white-box modeling   245
workby   105

X

XML (Extensible Markup
    Language)   69–72
XML schema definition (XSD)   69
XSLT (Extensible Stylesheet Language
    Transformation)   69

Z

Ziegler   259